Louis Arthur Coolidge, John F Pratt

Klondike and the Yukon Country

A description of our Alaskan land of gold from the latest official and scientific

sources and personal observation

Louis Arthur Coolidge, John F Pratt

Klondike and the Yukon Country
A description of our Alaskan land of gold from the latest official and scientific sources and personal observation

ISBN/EAN: 9783337418540

Printed in Europe, USA, Canada, Australia, Japan

Cover: Foto ©berggeist007 / pixelio.de

More available books at **www.hansebooks.com**

KLONDIKE

AND THE YUKON COUNTRY

A DESCRIPTION OF OUR

ALASKAN LAND OF GOLD

FROM THE LATEST OFFICIAL AND SCIENTIFIC SOURCES
AND PERSONAL OBSERVATION

BY

L. A. COOLIDGE

WITH A CHAPTER BY JOHN F. PRATT

CHIEF OF THE ALASKAN BOUNDARY EXPEDITION OF 1894

NEW MAPS AND PHOTOGRAPHIC ILLUSTRATIONS

PHILADELPHIA
HENRY ALTEMUS
1897

Copyrighted by Henry Altemus, of Philadelphia, in the State of Pennsylvania, A. D. 1897, in the One Hundred and Twenty-first Year of the Independence of the United States of America.

HENRY ALTEMUS, MANUFACTURER,
PHILADELPHIA.

INTRODUCTION

The object of this book is to furnish the latest authentic information concerning a portion of our country which until very recently has been little thought of; but which is now the magnet for many minds. The author wishes to acknowledge his indebtedness to officers of the U. S. Coast and Geodetic Survey, and the U. S. Geological Survey for helpful suggestions and for recitals of personal experience. He is especially under obligation to Mr. John F. Pratt, of the Coast and Geodetic Survey, whose service as Chief of the Alaskan Boundary Expedition of 1894 gives a peculiar interest and value to the chapter kindly contributed by him.

CONTENTS.

New Lands of Gold 7
Klondike and Yukon Diggings 22
Seeking the Pot of Gold 37
Life in Camp 63
Mining Experts and Scientists 78
Placer Mining and Hydraulics 94
Alaska 110
Quartz Mining in Southeastern Alaska 132
The Wonderful Yukon Country 144
The Boundary Dispute 175
Gold Production of the World 182
Our Northwestern Possessions, ... 185
Laws Governing the Location of Claims . . . 194
Climate of Alaska 208

KLONDIKE

CHAPTER I.

NEW LANDS OF GOLD.

On Wednesday, July 14th, 1897, the little steamer Excelsior arrived in the harbor of San Francisco with forty miners on board, each one of whom had brought with him from the ice-bound interior of Alaska a fortune in gold. From that day may be said to date the Klondike gold craze which already rivals in extent the three other great gold crazes of the century, California in 1849, Australia in 1851 and South Africa in 1890. Already the amount known to have been brought back by the returning miners exceeds $1,000,000, and nearly $3,000,000 more is said to be on the way. It is estimated by some experts that before the full returns come in it will be found the total output of the Alaskan mines has been $8,000,000. California yielded $60,000,000 five years after Marshall's discovery, and all from place diggings, as are the diggings in the Klondike region; but the facilities for mining in California, with its salubrious climate, its comparative nearness to civilization, its all-year-round conveniences, were infinitely superior to the facilities in the Yukon Basin, where winter lasts for ten months in the year, and where the thermometer drops to 72 degrees below zero in the winter and climbs to 120 degrees above zero in the summer, and where the nearness of the Arctic circle practically divides the year into one long day and one long night, each extending over a period of six months.

When millions of gold can be taken out in a single year under all these disadvantages of climate by laborers working with the most primitive implements of mining life it is difficult to conceive of the opulence of a soil whose grudging tribute to the energy of the modern argonaut is so fabulous in extent.

These forty men who came down on the Excelsior from the port of St. Michael, near the mouth of the Yukon, had among them over half

THE FIRST ARRIVAL AT DAWSON CITY.

a million dollars in gold dust, ranging in size from a hazel nut to fine bird shot and kernels of sand. All of them were penniless, or nearly so, when they left the United States, some of them having taken their departure within a year, others having been prospecting on the fields along the branches of the Upper Yukon for several years. They brought back fortunes ranging from $5000 to $90,000 and the most extraordinary tales of their experience in the mining countries. Their descriptions of the vast amounts of gold still remaining in the regions from which they had come were so tempered with cautions and warnings against a mad rush for the new fields that tales which otherwise might have been deemed improbable gained credence through their very conservatism. But whatever might be thought of the tales, there was no disputing the tangible fact of the yellow metal which was laid down in Selby's smelting works at San Francisco, and when a second ship, the Portland, from St. Michael, arrived at Seattle, three days later, with more miners aboard and $700,000 in bullion, it was as if a spark had set afire the enthusiasm for hunting gold which had been lying dormant since the days of the Argonauts of 1849. There have been few scenes in mining history more striking than that which was enacted when the men landed from the Excelsior, weather beaten, roughly dressed, with scraggly beards and furrowed cheeks, and marching straight to the smelting works, proceeded to produce bags of gold, dirty and worn, containing thousands of dollars in the precious metal.

As fast as the bags were weighed they were ripped open with a knife and the contents were allowed to scatter over the counter; and then some of the miners produced from bundles and coat pockets gold dust in all sorts of queer receptacles, such as fruit jars and jelly tumblers, and even writing paper, carefully secured with twine. No wonder the spectotors looked on with fascinated amazement. No wonder the strange news spread like wildfire. The gold fever of 1897 had begun to burn.

These miners brought the news that the new Eldorado was situated on the Klondike River, nearly two thousand miles from the mouth of the Yukon, just escaping the Arctic circle by a bare 250 miles, and situated in Canadian territory, a meagre 140 miles east of the 141st degree of longitude, which constitutes the boundary between Alaska and British America.

They told, too, of the terrible hardship through which they had gone in order to reach these marvelous gold fields and uncover their hidden wealth. Joseph Ladue, who left Plattsburg, N. Y., a few years ago, an impecunious farm hand, too poor to marry the woman of his choice, described how he had forced his way into the new diggings, established the city of Dawson, which is the metropolis of the gold region, and come back with thousands of dollars in hand and millions in prospect. But his most emphatic words were words of warning against those who would rush madly to the new field without considering the hardships they would have to undergo. Starvation and want, he said, would be the lot of those who ventured into the new Eldorado without a supply of provisions sufficient to last for months, and he said that those who ventured to leave for the North as late as August 1 were wasting their time, besides subjecting themselves to needless peril, for by the time they had traversed the long stretch of inhospitable country they would find winter setting in with Arctic vigor and they would be shut up in an ice-bound region hundreds of miles from telegraph or postoffice, a prey to starvation and cold.

Dawson City, which had sprung up in an Arctic night, was situated, they said, near the junction of the Klondike and Yukon Rivers, had a population when the miners left of 3,500, was laid out on modern lines with sixty-foot avenues and fifty-foot streets and had all the ambitious scope of a bonanza town with a few score log cabins and innumerable tents.

While the voyagers on the Excelsior were still telling their marvelous stories in San Francisco fuel was added to the fire by the arrival at Seattle of the steamer Portland, also straight from St. Michael, with sixty miners aboard and over $700,000 in gold. The excitement aroused by the arrival of the Portland surpassed even that of the earlier arrival, and it had hardly touched the wharf before hundreds of men in Seattle were crowding over one another to get an opportunity to board her for her return trip to the mouth of the Yukon.

These miners had been hunting for gold in the Yukon country for years. Some of them had found it in generous quantities lying in the streams and in the beds of creeks flowing into the Yukon just west of the spot where the river crosses the boundary between Alaska and Brit-

LYNN CANAL AT ENTRANCE TO CHILKOOT INLET.

ish America—along Forty-mile Creek, Sixty-mile Creek and Birch Creek. They would have continued digging along these creeks for months longer content with the moderate but certain returns of their labors had it not been for the sudden discovery on the Klondike pouring into the Yukon over on the British side, of gold nuggets so large and handily found that, carried away with the news, they pulled up stakes and abandoned in a day the claims upon which they had been toiling for months. Circle City, the largest camp in the Yukon district, was deserted over night, and Dawson City, at the junction of the Klondike and the Yukon, sprang into being in a day. This was a year ago, at the beginning of the short summer season. The gold the returning miners brought to San Francisco and Seattle was the product of that summer's pickings. They worked the Klondike and the banks of two creeks flowing into it, which they called appropriately the Eldorado and the Bonanza, until winter shut in on them, and for nine months of the cheerless Arctic season they lay huddled over their gold, until the breaking up of the ice in the following June gave them their first chance to escape back home with their treasure. They had been shut out from the world for nine months as completely as if they had been dead. They did not even know the result of the election for President. They were strangers in their own country.

The Portus B. Weare is a little steamer, owned by a transportation company, which makes the trip up and down the Yukon three or four times every summer, and on this boat the miners loaded their gold and left their fortune-banks behind. They steamed 2000 miles down the river to the diminutive port of St. Michael, on the coast of Behring Sea, there to take passage on steamers bound for home.

St. Michael is situated on an island ninety miles north of the mouth of the Yukon. It is the most important station of the coast for all the Yukon region, and, in fact, the only one so far as freight and supplies are concerned. On June 27, at noon, the Portus B. Weare, the first passenger steamer to arrive from up the river, came steaming around the low headland and drowned the frantic cheering of the crowds on the two boats lying there with its hoarse whistle. The Portland and Excelsior, drawing in excess of nineteen feet of water, were obliged to lie out a mile or more from shore, but the Weare, built for river traffic and drawing only a few feet, was enabled to steam up the shallow harbor and touch the dock. As she steamed near, friends who had not met in months or years greeted one another from deck to deck, and wives and children who had come to meet fathers and husbands, frantically threw kisses and wept and laughed by turns. A more exciting throng was never seen.

That the Weare brought good news was evident. Husbands, fathers and friends held up nuggets of glittering gold and bags of it before the eyes of those aboard the Portland, and the news was shouted across that a great strike had been made. "Circle City is busted!" "Only three white men left in it!" "The Klondike is the richest mining region on earth to-day!" "Hurrah for the new proposition!" "Circle City is the silent city!" These and kindred shouts rent the air. There was as great desire on the Portland to hear the news from up the river as there had been at St. Michael to hear from the outer world.

Those who were first to board the boat soon heard enough to convince them that on the Eldorado and Bonanza Creeks, branches of the Klondike, the richest strike in all American mining history, had been made. All the people knew was that gold had been found in such quantities that it seemed beyond belief; that all who went into the streams mentioned found gold, and that most of them or their partners were coming out and had gold to show. The Weare brought down on her first trip over $1,000,000 . Many of the men would not talk, but, with grips, bags, strong boxes, belts, tin tomato cans and other odd receptacles filled with the glittering metal, sat on guard in their 4x6 staterooms.

The purser was treasurer of the smaller holders. For Stanley and Worden he had $20,000; R. McNulty, $20,000; Henry Anderson, $20,000; C. D. Myers, $6000; T. Moran, $13,000; Joe Cozlies, $17,000; N. E. Pickett, $20,000; Victor Lord, $3500; C. A. Brannon, $7000; Albert Gray, $6000; N. Murcer, $15,000; John R. Moffett, $9000; C. H. Loveland, $8500; J. J. Hatterman, $12,500. Other men had sums far in excess of these, and, while some of them had given the purser from $5000 to $20,000 each to keep for them, retaining from $30,000 to $100,000 themselves, others had retained all. Some of the following are among those who had treasure with the purser:

Clarence Berry, $110,000; Henry Anderson, $65,000; William Stanley, $112,000; J. Clements, $50,000; Frank Keeler, $50,000; T. J. Kelly, $33,000. The following men had from $30,000 to $100,000 each: Frank Phiscater, Nat Hall, A. McKenzie, B. F. Purcell, O. Finstead, Charles Silverlock, Jeremiah Johnson, Pete Copeland, C. E. Myers, F. Bellinger, R. H. Blake, Joe Burgoyne, William Sims, John J. Moffatt, Joe Debosher, Fred Tabler, William Sloan, C. H. Loveland, N. Mercer, Charles Emcher, Harry Oleson, Charles Anderson, Henry Plato, Honora Gotthier and John Williamson.

Most of the sixty passengers aboard the Weare, which started from winter quarters after the ice started in the Yukon, had been living on beans, bacon and bread or hard tack for from six months to a year, some longer. The little agency store at St. Michael was besieged for bottled cider, canned pineapples, apricots, cherries, or anything tart, and at a dollar a bottle cider went like mad. They were eager for raw turnips, and even for potatoes, and when a crate of onions was sent over to the Weare from the Portland there was almost a riot to get at them.

The richest gold strike the world has ever known was made in the Klondike region last August and September, but the news did not get even to Circle City until December 15, when there was a stampede. Circle City was deserted. But three white men and several Indians and women came out to greet the returning miners as they came down stream.

George Carmack made the first great strike on Bonanza Creek August 12, and on August 19 even claims were filed in that region. Word got to Forty Mile and Circle City, but the news was looked on as a grub-stake rumor.

December 15, however, authentic news was carried to Circle City by J. M. Wilson, of the Alaska Commercial Company, and Thomas O'Brien, a trader. They carried not only news, but prospects, and the stampede was on. Those who made the 300-mile journey the quickest truck it the richest. Of all the 200 claims staked out of the Bonanza and Eldorado it is said not one proved a blank, and it was learned as the Weare left the diggings that equally rich finds had been made on June 6 to 10 on Dominion

Creek. This last creek heads at Hunker Creek and runs into Indian Creek, and both run into the Klondike. Three hundred claims have already been staked out on this Indian Creek, and the surface indications show that they are as rich as any of the others.

The largest nugget yet found was picked out by Burt Hudson on Claim Six of the Bonanza, and is worth a little over $250. The next largest was found by J. Clements, and was worth $231. The last four pans Clements took out ran $2000, or on an average of $500 each, and one of them went $775. Bigger pockets have been struck in the Caribou region and in California, but no where else on earth have men picked up so much gold in so short a time. A young man named Beecher came down a-foot and by dog sledge, starting out early in March. He brought $12,000 to $15,000 with him. He was purser of the Weare last summer, and went in after the close of navigation in October or September. About December 15 he got a chance to work a shift on shares, and in sixty days made his stake, which was about $40,000. Gold is in circulation in Dawson in fabulous amounts. Saloons take in $3000 to $4000 each per night. Men who have been in all parts of the world where gold is mined say they never saw such quantities taken in so short a time.

At least $2,500,000 has been taken from the ground on the British side within the past year, and about $1,000,000 from the American side. The diggings around Circle City and in the older places are rich.

There was one woman in the throng of miners who came from the Yukon on the steamer Portland. This was Mrs. J. S. Lippy, the wife of Prof. Lippy, who a year or two ago was secretary of the Y. M. C. A. at Seattle and who brought back with him $85,000 in gold. Mrs. Lippy was the first white woman on the creek and the only one in her camp, but she was not the first white woman to cross the divide. Nine or ten others were at Forty-mile Creek.

Lippy went to the gold fields with hardly a grub stake. He believes his claim is worth $350,000. It may be worth millions.

Joseph Ladue, formerly of Binghamton, N. Y., was a farm hand before he went to Alaska. He struck it rich and is the owner of the town site of Dawson City. He counts himself a millionaire. He went to the Northwestern country first in 1892 and has been there most of the time since. He left Dawson with a population of 3500. He was the first man to run a saw mill in Alaska, and it was a paying investment, although it was almost impossible to get anybody to run it. He paid men as high as $15 a day to work for him. The cheapest lumber he ever sold brought $100 per thousand, and when planed double that amount. Mr. Ladue, since his return, has said that already eight hundred claims are staked within a radius of twenty miles of Dawson. There is jumping of claims. Three months' work each year is required to hold a claim. Failing in this the land reverts to the government. The laws of Canada are stringent in such matters, and severe penalties are imposed for jumping or other interference with the rights of claimants.

Another successful argonaut is William Stanley, 68 years old, who up to two years ago kept a little stationery stand in Seattle. He left a wife at home with several children and took one son to the gold fields with him. He brought back $112,000 and left his son in the diggings. He is interested with his son and two New York men in claims which he values at $2,000,000. He went to the Yukon as a last resort, and made his findings in three months.

Ethel Bush, of Selma, Cal., and Clarence Berry, of Fresno, were married March 15, 1896. They were penniless, and for a honeymoon they chose a journey to the Alaskan gold fields. They drove their dog team into Forty Mile camp eighty-seven days later. For weeks they toiled on without result. Then came the Klondike find, and they moved on to Dawson City, where they picked out over $100,000, and they sold their claim in San Francisco for $2,000,000.

CHAPTER II.

THE KLONDIKE AND THE YUKON DIGGINGS.

The richest yields of gold in the Yukon region have come from the territory embraced by the 138th and 145th degrees of longitude and the 62d and 66th degrees of latitude, between the upper ramparts on the East—steep bluffs frowning on a picuresque bend in the river, and Fort Yukon on the west. The greatest extent of gold-

IN CAMP NEAR FORTY MILE CREEK.

KLONDIKE AND YUKON DIGGINGS.

bearing territory thus far explored is the American side of the 141st degree of longitude, which is the accepted boundary line. The most sensational discoveries have been on the British side, about 140 miles to the east of the line.

On the American side gold has been found in liberal quantities along a number of creeks, Birch Creek, Firty-mile Creek and Sixty-mile Creek being the most promising fields in the order named, and the centre for these diggings has been Circle City, on the bank of the Yukon, about 140 miles west of the boundary. On the British side of the Klondike River and the Eldorado and Bonanza Creeks, tributary to it near its junction with the Yukon, have proved the miners' paradise. There is a group of creeks very near the boundary, chief of which is Miller Creek, which have contributed most generously to the gold supply. These are claimed both by the American and the British officials, and there is grave danger that they may lead to international complications unless the boundary is quickly surveyed.

Miller Creek, up to the time of the discovery of Klondike, was credited with the richest diggings along the Yukon in proportion to their extent. Over $300,000 was taken out last season. The creek is only six miles long, but fifty-four claims were staked out on it. A claim consists of 500 feet of creek and reaching up indefinitely on both sides of the gulch. The creek is distant about sixty miles from Forty-mile Post, at the junction of Forty-mile Creek with the Yukon, and it is surrounded at short distances by Poker, Davis, Glacier and Little Gold Creeks, all bearing gold.

The Klondike River enters the Yukon from the east at a bend about 300 miles east of Circle City and fifty miles north of Sixty-mile Creek. From Sixty-mile Creek the course of the Yukon is due north to the Klondike and then it starts again toward the West. The great copper belt crosses the Yukon just at this point, and the Indians have had a fishing camp there for years, the Klondike being a noted stream for salmon. Its waters are very clear and shallow, as befits its source high up in the snow-capped ranges.

"Klondike" means "reindeer." It is about as near the Indian word as the miscellaneous population of prospectors who have been digging there for gold were able to come. At the United States Coast and Geodetic Survey it is said the

KLONDIKE AND YUKON DIGGINGS. 25

word ought really to be spelled "Tlondak," which is Indian for "fishing grounds," and that is the name given to the stream which has now become synonymous with Eldorado in maps which were made in 1887 by Mr. McGrath, the Coast Survey official detailed at that time to explore a country which was then quite unknown. McGrath very nearly starved to death on the very spot whence millions of dollars in yellow metal have been taken during the last twelve months, and he never suspected the presence at that immediate place of the precious metal. But that is another story.

Miners have been taking out gold since 1894 from the placer diggings on the American side of the line. The earliest diggings were at Forty Mile Creek, about sixty miles east of the Klondike, and then came discoveries at Sixty Mile Creek, a little farther south, and at Birch Creek, a good deal farther west. Of these diggings those along Birch Creek have been the most profitable, and the camp of Circle City, which was founded in the fall of 1894, was for a time a place of considerable importance. It was the distributing point for the whole region and was, in a measure, the metropolis of the Yukon Valley. Now it has been eclipsed, for a time, at any rate, by the new settlement at Dawson City. Circle City has the great advantage, however, of being on American soil, for whatever the present temporary tendency, it is believed by those who have studied the country most closely that the American side of the 141st parallel of longitude, which constitutes the Alaskan boundary, will eventually prove the richest and most profitable portion of the gold-bearing territory. Over 500 men wintered at Circle City last year. The town, which is situated near the head waters of the Yukon, about 170 miles from Forty Mile Creek, is laid off in streets, with the main street facing the river, and it is so near to Birch Creek that a portage of six miles brings it to the banks of Birch Creek, two hundred miles from the mouth, and thus in a position to bring the gold ores taken out of this great American gold-bearing basin to the navigable waters of the Yukon. The gold diggings on American soil which have been prospected extend from the 141st to the 146th degree of longitude. The Klondike region is just to the west of the 141st degree, Dawson City being situated at the junction of the Klondike and Yukon, about sixty miles to the west.

KLONDIKE AND YUKON DIGGINGS. 27

The experts of the Coast and Geological Surveys who have explored the country to some extent estimate that the gold-yielding territory extends over at least five hundred miles and that the richest portion of it is on American soil. The Cassiar Mountain region, as far east as the 130th degree of longitude on the northern border of British Columbia, has been worked with a good deal of success for the last eleven years, although the yield now seems to be falling off. The gold in this region comes from the same mother lode as that at Klondike, at Sixty Mile Creek, at Forty Mile Creek and at Birch Creek. Scientists believe it is from the same mother lode as the gold from the Sierras, and they even go so far as to assert that the gold mines of the Ural Mountains in Siberia go back to the same origin. In other words, the whole country of two continents, from the Ural Mountains to the Rockies, is impregnated with a mineral which is apparently exhaustless in extent and which will suffice to keep the world supplied with gold for ages to come.

Nobody seems to know just when gold was first discovered in the Yukon Basin, for no two miners can be found to agree on the subject. It seems to be certain that none was ever found there before 1860, although it is said that some of the Hudson Bay Company's men ran on to gold at about that time. But if they did the discovery was never followed up, and they are hardly entitled to the credit. It does not appear that the Russians, during their ownership and occupation of the country, ever instituted any thorouh search for the precious metals. It is true that gold was discovered by Doroshin on the Kenai Peninsula in 1848, and that he afterwards, in 1850-51, made further explorations of the same neihborhood, but it has always been charged that the Russian-American Company, regarding, as it did, any effort to develop the mineral resources of the country as in the highest degree inimical to the business in which it was wholly engaged and of which it held an exclusive monopoly, induced him, by the payment of a consideration, to suppress the truth in regard to what he may really have discovered. There is a tradition, too, among old Russian residents that a Russian engineer sent out by the Imperial Government to examine and report on the mineral resources of the country, made some rich discoveries on Baranoff Island, which he reported in Sitka, whereupon, being of convivial habits, he was taken in charge by the governor, who was also the company's manager, by whom he was wined and dined and his appetite for drink ministered to until he sank into a drunkard's grave, and was thus prevented from making any report of his discoveries to the Imperial Government. Doroshin did, however, report finding gold on the Kaknu River, which empties into Cook's Inlet, though it appears that his explorations were wholly confined to an examination of the alluvial sands of the streams and gulches in that neighborhood. To the fact that the Russian-American Company, like the Hudson Bay and American Fur Companies, believed that its interests would be jeopardized by the bringing to light of any natural resources which would invite immigration, and thus tend to the early settlement and development of the country, is no doubt due the further fact that nothing was publicly known before the transfer of the existence in Alaska of gold and silver in paying quantities.

So far as is known, the first genuine prospector in the Yukon region was one George Holt, who is declared to have been the first white man to cross the coast range for that purpose. About all that is known of Holt is that he made his journey in 1878, but nobody seems to know what path he followed or whether he took the trail over the Chilkoot or White Pass. It is known only that he descended the chain of lakes above the Chilkoot Pass, which have since been traversed by so many other seekers after gold, that he followed the Indian trail to the Hootalinqua River and that he returned the same way in the fall. The Hootalinqua River region, which he penetrated, is about two hundred and fifty miles to the southwest of the Klondike. Holt reported that he found coarse gold near there, but no coarse gold has been discovered in that region since, although flour gold has been yielded up from the bars of the river. In any event, Holt did not find encouragement enough to continue his exploration. The next that is known is the expedition of Edward Bean, who started out from Sitka in 1880 at the head of a prospecting party. There were twenty-five men in the company. They crossed Chilkoot Pass to Lake Lindemann, built boats and descended the Lewis River as far as the Hootalinqua. Their success amounted to the finding of gold in a small stream fifteen miles above the canon yield-

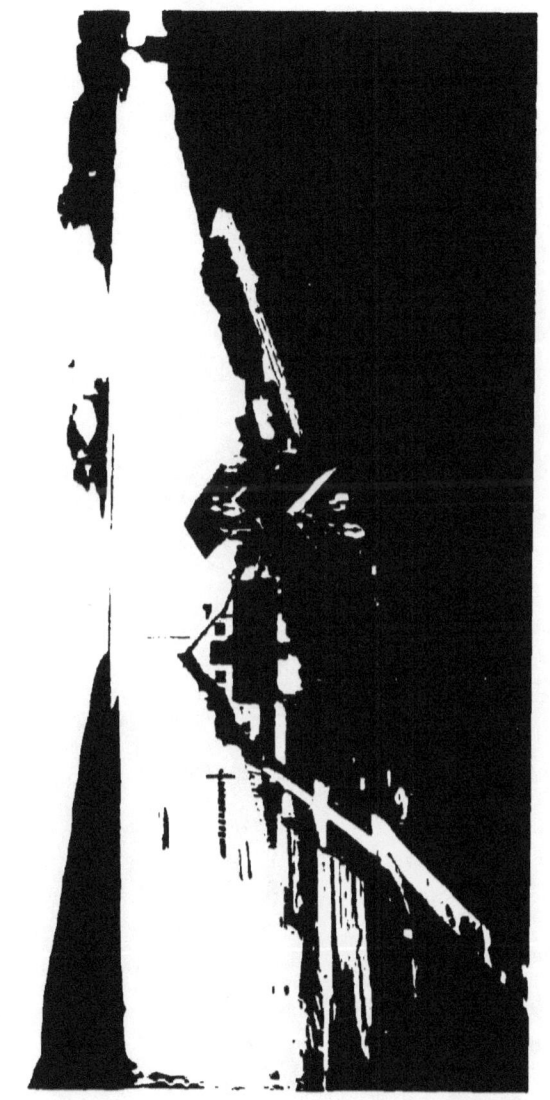

THE LANDING AT DYEA.

ing $2.15 per day. This was not a discovery calculated to encourage further attempts, but about this time many other small parties began to force their way through the Chilkoot Pass farther and farther up the lakes and the rivers. All of them found gold in greater or less quantities. The first party to discover gold in really paying quantities in the Yukon Basin consisted of four miners, who crossed the range in 1881 and descended the Lewis River as far as the Big Salmon River, ascending that stream for over two hundred miles and finding gold on all its bars. The Cassiar Bar was not located until 1886, and up until a comparatively recent time this was the richest of all the bars ever located on the Yukon or any of its tributaries. It was in the same year that coarse gold was found on Forty Mile Creek on American soil several hundred miles down the river. This discovery drew off all the miners who had been digging in the upper river country on Canadian territory. The bars at Forty Mile Cretk were worked for some years at a good profit, but they have now been abandoned owing to the discovery of coarse gold more easily accessible in the gulches. Forty Mile Creek, which will always be of interest from the fact that it was the scene of the first touch of gold excitement in Alaska, owes its name to the fact that it enetrs the Yukon about forty miles from Old Fort Reliance. It is about two hundred and fifty miles long and has many tributaries, all of which carry gold in paying quantities. Sixty Mile Creek enters the Yukon River from the southwest and about seventy miles above the mouth of the Stewart. It has given up excellent yields of gold, and about 100 miners have wintered every year of late at a trading post and a saw mill which have been established on one of its islands Birch Creek was not prospected until 1893, and then only just enough to show that the country contains some gold. In the season of 1894 nearly one hundred men prospected this country and staked off their claims. It was found that bedrock was much nearer the surface than in the Forty Mile Creek district, and the claims yielded very good returns. They drew many men away from the Forty Mile Creek mine.

The mining of these regions is still in its infancy, although it has been going on in more or less desultory fashion for the last fifteen years, and only a few of the most accessible streams have ever been prospected. All the larger rivers of the upper country furnish flour gold which increases in coarseness as the rivers are ascended, and from this it is argued that the surrounding gulches in many places must furnish exceedingly rich diggings. The territory cut by these streams has never been explored even superficially except as it may have been explored in the last year by miners hunting for gold, and yet it is almost unlimited in extent. A hundred thousand men could be hunting gold in the Yukon Basin at the same time without ever crossing one another's tracks and each would be lost to the world.

The honor of discovering the richest placer mines in the world belongs to an Illinois man named George Carmack. A party of miners, drifting by the mouth of the Klondike on July 9, 1896, found Carmack camped in a lonely spot there with his family. His wife was a native woman of the Stick tribe, and he had two dark-skinned children following him about.

He had been in the country eight years, and much of the time had been spent with the Sticks at Tagish House, on the chain of lakes that form the source of the coastward arm of the Yukon, on the trail from Juneau to the gold fields. When found Carmack was making quite extensive preparations for curing salmon, the annual run of which was expected to begin any day. He had erected a birch-covered shed for the better protection of his catch from the weather, and he already had his nets at the mouth of the Klondike, a half-mile farther down. Carmack expected to sell his crop the following winter, principally for dog feed, although in times of food famine, as really occurred last winter, dried salmon became a staple article of diet for white men.

Carmack told his visitors of his intention to prospect the Klondike as soon as the salmon season was over. Four weeks later he took two Indians and started up the stream. After a few miles of laborious poling against a rapid current they turned into the first considerable tributary that came in from the right. Here conditions were favorable for prospecting, the water being shallow, and they found gold in encouraging quantities on the bars of the creek. They followed the windings of this stream for twenty or twenty-five miles before they made locations and went to work.

The results were almost enough to turn the

brain of a prospector who had searched for many years in the hope of finding gravel that would yield a few grains' weight of gold to the pan. Here at a depth of three feet in the low bars by the creek they found dirt that carried a dollar to the pound in coarse, ragged bits of gold. Others have since found diggings ten-fold richer.

As remote as their discovery was they were not long to remain in sole possession of it. With the exhaustion of their few days' provisions, the two Indians were sent back to the village for supplies.

About the middle of August, when the P. B. Weare, on one of its occasional trips, arrived at the Indian village, which is about half way between Forty Mile and Sixty Mile Creeks, the Indians were waiting there to lay in their supplies. There were also several other prospectors who happened along, and the discovery was now common talk. The stories of fortune proved a little too much for the crew of the Weare to withstand. They deserted in a body and joined the rush to the new gold fields. The captain, after being delayed three or four days, got an Indian crew sufficiently trained to handle the boat. When he arrived at Forty Mile on his return the reports were alluring enough to impel a hundred or more men to start at once for the new find.

The Klondike had been known for several years to drain a gold country, and the first five miles of it had been indifferently prospected, but the gold hunters were generally run out by bears.

If the miners had made any encouraging finds at the outset it would have been different, but all other things being equal, in their estimation, they concluded to try streams where the bears were not so aggressive. And it happened that there was a reason for the bears being so bad in that particular place. It is possibly the best stream for salmon of all the tributaries of the great river.

The mountains along that section of the Yukon, and in fact, from Circle City up stream for several hundred miles, are extremely wild and rugged. The great copper belt, which crosses the Yukon at the Klondike, is a succession of massive quartz ledges, with that metal predominating. The veins are known to carry gold, but in what proportion is yet to be determined.

Here also is the pioneer quartz mine of the Yukon. Captain Healy, the manager of the transportation company, located a claim on the side of a precipice opposite the mouth of the Klondike over two years ago. Vein mining had never been thought of as a present undertaking. Labor was worth $15 a day and supplies of all kinds were proportionately high, but he put up his location.

Last year he did some development work on it and had samples assayed, showing it to be rich in gold. But the latest reports from the Klondike put such extravagant prices on labor that quartz will not be considered for some time yet.

Still it is in the veins that will be found the real wealth of this wonderful country.

CHAPTER III.

SEEKING THE POT OF GOLD.

The first requirement for one seeking the gold fields is a hardy constitution; the second is capital. For the Yukon is not, as some other gold countries have been, a poor man's paradise. Gold is there in Aladdin-like profusion, but it is not to be had for the asking. It comes only as the fruit of wearisome and perilous travel, of desperate combat with the rigors of an Arctic climate, of deadly waiting for Arctic winters to unloose their icy hands. For the privilege of a few months of toil the prospecting miner must endure many months of unremunerative delay, during which he must pay extortionately for the mere privilege of living. For the season of placer mining lasts only during June, July and August.

Before beginning even to hunt for gold the aspiring miner must prepare himself for the long and tedious trip to the fields, and this is a task that will tax the endurance and nerve of the most hardy. It means, according to one who has made the trip, "packing provisions over pathless mountains, towing a heavy boat against a five to an eight-mile current, over battered boulders, digging in the bottomless frost, sleeping where night overtakes, fighting gnats and mosquitoes by the millions, shooting seething canyons and rapids and enduring for seven long months a relentless cold which never rises above zero and frequently falls to 80 below."

Any man who is physically able to endure all this, who will go to the gold fields for a few

ICE FORMATION AT WHITE HORSE RAPIDS.

years, can, by strict attention to business, make a good strike, with the possibilities of a fortune.

But he must have money to start with. All who have been to the gold fields agree in saying that no man should undertake the journey with less than $400 in capital. And he had better have $1000. The expense of reaching the mines is considerable. One hundred and fifty dollars is a modest figure for the journey from Seattle, and when once in the gold region the expense of living is enormous. The prices of even the most ordinary provisions are fabulous, and the companies doing business there refuse to give credit, as they can sell all their goods and more for ready cash. Provisions are almost unobtainable at any price. An officer of the U. S. Geological Survey, who has traveled through this country, has assured the author of this book that if he were looking for certain profit and had the necessary capital he would never think of hunting for gold, but would invest everything in provisions and groceries, which would yield enormous profits should they be got into the Yukon region.

If the traveler contemplates the overland trip his outfit should be bought in Juneau, the metropolis of Southeastern Alaska, the last outpost of civilization in the path of the voyager for gold. The needs of the traveler can be gauged there better than anywhere else, nearer the centre of population and wealth. Experienced men have found that the provisions a man ought to lay by before starting on the overland journey from Juneau make a formidable list. The articles required for one man for one month are somewhat as follows:

Twenty pounds of flour, with baking powder.
12 pounds of bacon.
6 pounds of beans.
5 pounds of dried fruits.
3 pounds of dessicated vegetables.
4 pounds of butter.
5 pounds of sugar.
4 cans of milk.
1 pound of tea.
3 pounds of coffee.
2 pounds of salt.
Five pounds of corn meal.
Pepper.
Matches.
Mustard.
Cooking utensils and dishes.
Frying pan.

Water kettle.
Tent.
Yukon stove.
Two pairs good blankets.
One rubber blanket.
Bean pot.
Two plates.
Drinking cup.
Tea pot.
Knife and fork.
Large cooking pan.
Small cooking pan.

These are simply for sustenance. In addition the traveler will find it necessary to build his own boat with which to thread the chain of lakes and rivers leading to the gold basin. He will need the following tools:

Jack plane.
Whip saw.
Hand saw.
Rip saw.
Draw knife.
Ax
Hatchet.
Pocket rule.
Six pounds of assorted nails.

Three pounds of oakum.
Five pounds of pitch.
Five pounds of five-eighths rope.

He will also find that he must have some protection against the deadly assaults of gnats and mosquitos, which fill the air throughout Alaska; that he will have to be provided for mountain climbing and for protection against snow blindness, whch is one of the most demoralizing afflictions that can befall the traveler over the snow-covered passes. So he will need:

Mosquito netting.
One pair crag-proof hip boots.
Snow glasses.
Medicines.

These are the provisions necessary for a miner for a single month, and whether he will need more for his journey depends somewhat upon the manner in which he travels. In the first place nobody should undertake to travel alone. The trip should be made in parties of two or more, which will conduce to safety and also lightness of the individual's load. It is possible for parties to attend to their own transportation over the divide between Juneau and the lakes. In that case they should start before the first of

April so as to catch the snows and ice. They can use sleighs over the summit of Chilkoot Pass and along the lakes down to the place of junction with the river. By the time the river is reached the ice will have begun to break away and the rest of the journey can be managed by boat. By this arrangement the gold fields can be reached four weeks earlier than by waiting for the opening of the summer season before starting from Juneau. Should the start be deferred till after April 30, Indians will have to be employed to do the packing across the pass. The Indians charge $14 per hundred for this service, and each is accustomed to carry about a hundred weight.

Before making a start the wise traveler will consider the cost of living in the diggings and provide himself accordingly. Following are a few of the average prices of provisions and articles of common use:

Cost of shirts	$5.00
Boots, per pair	10.00
Rubber boots, per pair	25.00
Caribou hams, each	40.00
Flour, per fifty pounds	20.00
Beef, per pound (fresh)	.50

44 KLONDIKE.

Bacon, per pound	.75
Coffee, per pound	1.00
Sugar, per pound	.50
Eggs, per dozen	2.00
Condensed milk, per can	1.00
Live dogs, per pound	2.00
Picks, each	15.00
Shovels, each	15.00
Wages, per day	15.00
Lumber, per 1000 feet	150.00

When the miners left Dawson City the following prices were in vogue:

Flour, per 100 lbs.	$12.00
Moose ham, per lb.	1.00
Caribou meat, per lb.	.65
Beans, per lb.	.10
Rice, per lb.	.25
Sugar, per lb.	.25
Bacon, per lb.	.40
Butter, per roll	1.50
Eggs, per dozen	1.50
Better eggs, per dozen	2.00
Salmon, each $1 to	1.50
Potatoes, per lb.	.25
Turnips, per lb.	.15
Tea, per lb.	1.00

Coffee, per lb. .... .50
Dried fruits, per lb. .... .35
Canned fruits .... .50
Canned meats .... .75
Lemons, each .... .20
Oranges, each .... .50
Tobacco, per lb. 1.50
Liquors, per drink .... .50
Shovels 2.50
Picks 5.00
Coal oil, per gallon 1.00
Overalls 1.50
Underwear, per suit .... $5 to 7.50
Shoes 5.00
Rubber boots .... $10 to 15.00

The tourist from the Atlantic seaboard will find in the following table information concerning the expenses of travel according to his means and inclination:

Fare from New York to Seattle via Northern Pacific, $81.50.

Fee for Pullman sleeper, $20.50.

Fee for tourist sleeper, run only west of St. Paulu, $5.

Meals served in dining car for entire trip, $16.

Meals are served at stations along the route a la carte.

Distance from New York to Seattle, 3290 miles.

Days required to make the journey, about six.

Fare for steamer from Seattle to Juneau, including cabin and meals, $32 cabin; $17 steerage.

Days, Seattle to Juneau, about five.

Number of miles from Seattle to Juneau, 725.

Cost of living in Juneau, about $3 a day.

Distance up Lynn Canal to Dyea, steamboat, 75 miles.

Number of days New York to Dyea, twelve.

Cost of complete outfit for overland journey, about $150.

Cost provisions for one year, $200.

Cost of dogs, sled and outfit, about $150.

Steamer leaves Seattle once a week.

Best time to start is early in the spring.

Total cost of trip New York to Klondike, about $667.

Number of days required for journey, New York to Klondike, thirty-six to forty.

Total distance, Juneau to the mines at Klondike, 650 miles.

Having settled the question of expense, the next thing is to select a route. The routes that go into Klondike are two. The most expensive

CHILKAT MOUNTAIN.

SEEKING THE POT OF GOLD. 47

is by steamer from Seattle to St. Michael, a distance of 2500 miles, and then by river boat up the Yukon, 1700 miles to Dawson City. By this route it takes thirty-five or forty days, and the fare is $180. The steamers permit only 150 pounds of baggage for each passenger. The two steamers that leave before the river is closed by ice this fall cannot carry more than 150 passengers each. This route is the more expensive, and some think the more comfortable.

The second route is overland from Juneau, and is the most perilous, the most subject to hardships and consequently the most fascinating fortune-hunting journey that could be imagined. Steamers run from Seattle to Juneau, which is the metropolis of Alaska, and thence a small steamer transports the seeker after gold up Lynn Canal and Chilkoot Inlet to Dyea, sometimes called Taiya, which has just been made a port of entry by Secretary Gage for the benefit of the incoming horde of miners. The distance is 650 miles. Dyea is just at the head of the northernmost branch of Chilkoot Inlet, which is itself a branch of Lynn Canal, the extreme northern limit of navigation, and is one hundred miles due north of Juneau. At Dyea the overland journey begins. The outfit, which for the long period of isolation in the interior is no small affair, is packed on sleds and hauled for twenty-seven miles over the mountains and over the deadly Chilkoot Pass to Lake Lindeman, the first of the series of lakes reaching up into the interior. This passage of twenty-seven miles is the most difficult part of the whole journey. It would be bad enough if it were made without baggage. A good traveler, in prime condition, unhampered by an elaborate outfit, can make the summit of Chilkoot Pass from Dyea in twelve hours. Mr. Pratt, of the United States Coast Survey, who was in Alaska on the boundary commission several years ago, left Dyea with a companion at 9 o'clock one morning and reached the summit of Chilkoot Pass at 9 o'clock the same night. But that was a case of moving light infantry. Ordinarily it will take a miner at least two days to make the difficult ascent with a portion of his outfit, and sometimes it is necessary for him to go back to the starting point for the rest of his outfit, for it is to be borne in mind that transportation companies have not yet secured a charter to do business in Chilkoot Pass. Thus it is that at least six days might be used up in getting over the short distance from salt water to fresh. Sometimes it takes even longer than that. The only assistance that can be obtained is that of the Indians, who can be hired to carry outfits over the divide at an expense of $14 for every hundred pounds. This is done in the absence of snow, which precludes sledding. With the present rush to the gold fields the natives will receive large profits. The overland trip involves a climb of 3500 feet to the summit of Chilkoot Pass, and it is one of the most impressively picturesque journeys that can be imagined. The landscape is resplendent with glaciers, the ice sparkles like jewels in the Alaskan sun, the mountains rise in the distance on every side, and it is all impressive beyond the power of description. Beyond this the trip is exhausting, though necessarily not so dangerous as in the pass, for there are times when sudden snows come to fill in the pass without warning, and there are few who have survived such an encounter with the elements as this. But with Lake Lindeman a new feature of the journey presents itself. Those who make the journey in summer will find the ice out of the lakes, but if an early start were to be made they would be able to cross Lake Lindman and the other lakes of the chain on foot or else by means of ice boats temporarily constructed. The ice breaks up in the lake about the first of May, and then it becomes necessary for the travelers to stop and build boats. Until the last year it was necessary for every miner to carry a whip saw with him with which to cut the timber for his craft, and whip-sawing was one of the picturesque, although not especially inviting, incidents of the trip. But a saw mill has recently been constructed. The only timber used in the construction of boats on the lakes is spruce or Norway pine. In the first place, the timber has to be located, and this is not the easiest thing in the world, because the timber around the lake is nearly all burned off, and there is none suitable for boat building. After the timber has been located comes the construction of a saw pit. To construct a saw pit it is necessary to find trees so arranged as to support cross-pieces, the stumps being cut at a proper distance from the ground so as to take the notched cross-pieces in. This requires four trees about equi-distant from one another, and the cross-pieces have to be fixed very firmly in place so as not to slip, as the log which is to be sawed is likely otherwise to be the cause of an accident. Often a good woodsman will be able to fell the tree which is to be sawed in such a way that it will fall into the pit, which saves the time and trouble of skidding the log up and rolling it in place after felling, which is frequently a very difficult task. From the slabs and boards thus roughly made the flat-boats are constructed, upon which the miners traverse the chain of lakes extending north from Chilkoot Pass. Lake Lindman is about six miles long, with an average width of one mile, and is cleared to navigation usually after May 15th, although sometimes not before June 10th. Connecting with it is Lake Bennett, which is twenty-six miles long, with an average breadth of one mile; and then comes Tagish Lake. Lake Bennett is surrounded by high mountains, which rise abruptly on either side, making it exceedingly difficult to find a landing place. It is rather perilous for rafts and boats on account of the strong winds which sweep up from the south through Chilkoot Pass, Lake Bennett acting as a funnel for that norrow passage. The winds are always in the south and are caused by the hot air of the inland valleys, supplemented by the cooler air of the coast, rushing inland over the low passes and down the lakes. As Lake Bennett is only five miles wide at its broadest place, and at many points is much less than a mile, the air is forced over it between the high ridges of mountains at a tremendous rate. Some of the mountains reach a height of 8000 feet. The climate is dry, and what little rain falls consists of an occasional thunder shower. The air is cool and bracing from the snow-capped peaks, which temper the warmth of a down-pouring sun.

Lake Bennett is connected with Lake Tagish by a very crooked and shallow channel with a slight current known as Caribou Crossing, from the fact that it was used by the bands of barren land caribou in their migrations in the fall and spring. Tagish Lake is an irregular body of water with two arms, known as Windy Arm and Taku Arm, stretching off to the south and southeast. Taku Arm is really a larger body of water than that particular portion known as Tagish Lake, but Tagish Lake acquires its importance from being directly in line of travel between Lake Lindeman and Lake Bennett on the south and Lake Marsh on the north. Tagish Lake is connected with Lake Marsh by a broad river with slow current, lined with wooded slopes and plenty of cottonwood and white spruce. The river is

SEEKING THE POT OF GOLD.

about five miles long, and on it is situated the Tagish House, where yearly festivals and councils of war are held by the natives, the buildings being the only permanent structures in hundreds of miles above where the Pelly and Lewis Rivers join to make the Yukon.

Lake Marsh, which is next entered, stretches along at a width of two miles for a distance of twenty, the most notable feature of all these lakes being their narrowness as compared with their length. Lake Marsh is in the middle of a broad valley, from which high ranges of mountains stand out prominently at a considerable distance. Its banks, like the banks of the other lakes, are well wooded. From Lake Marsh the seeker for gold finds his way into Lewis River, which he follows for a distance of more than a hundred miles to the northwest until he reaches the gold fields around the Klondike Basin. This journey along Lewis River, with its canons and rapids, is one of the most picturesque and interesting that can possibly be imagined. One of the features of the trip is the high cut banks which stretch along for mile after mile and which are completely honeycombed by martins, which resort there to rear their young. Lake Marsh is the limit for the migration of the salmon, which arrive there in small numbers, although those who do brave the journey are said to be the finest to be found anywhere in the world, averaging forty pounds in weight. The swift waters of the Grand Canon are too powerful ever for the salmon whose hardihood brings them as far up the river as this.

The Grand Canon is a wonderfully beautiful bit of scenery. It is cut through a horizontal basalt bed, and the walls range in height from fifty to one hundred and twenty feet, being worn into all sorts of fantastic shapes. The average width of the canon is about one hundred feet, and as the average width of the river above it is over seven hundred feet, the force with which this great volume of water cuts through the steep ledges of rock may be imagined. Mr. Wilson, who made this trip in 1894 and who has described it at length in his "Guide to the Yukon Gold Field," says that he shot through the canon for a distance of three-quarters of a mile in two minutes and twenty seconds, and when his boat emerged from the chasm it was leaking badly and nearly every nail was started. Two miles beyond come the White Horse Rapids, which form a perilous passage even for the best of

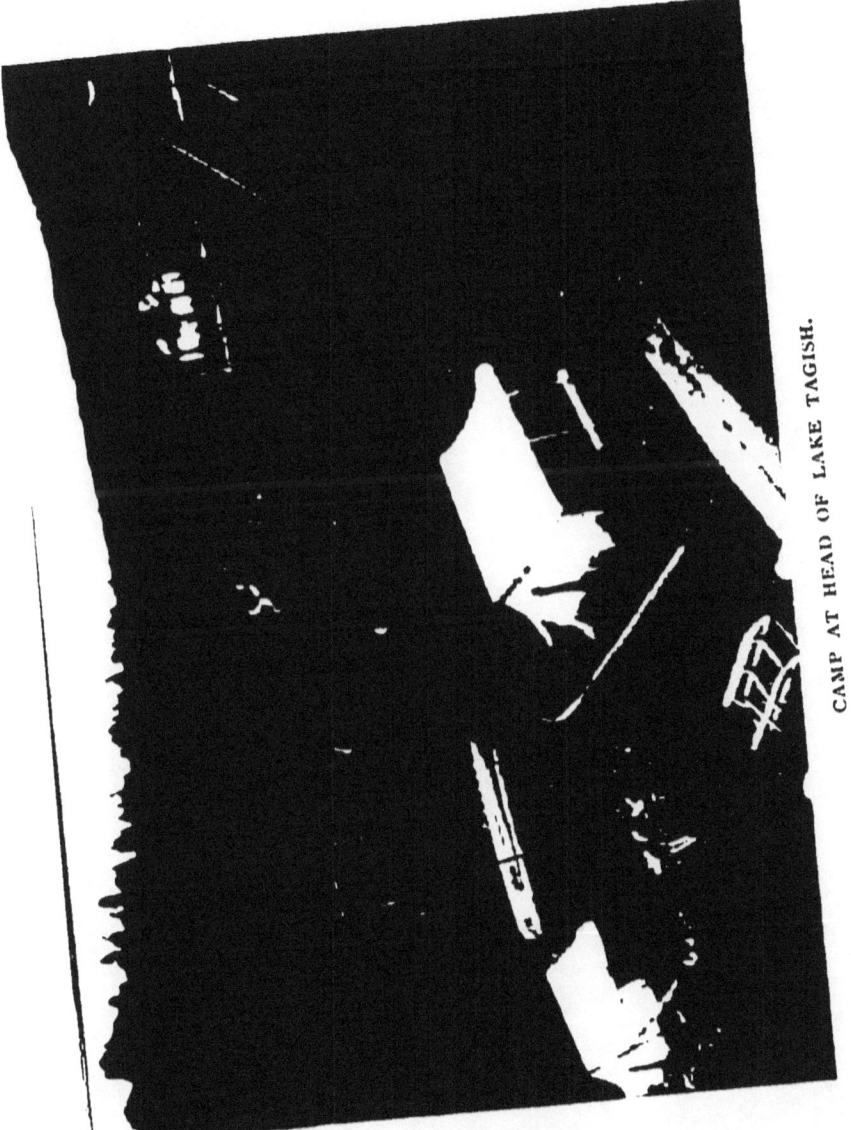

CAMP AT HEAD OF LAKE TAGISH.

boats, and farther down comes Lake Labarge, at a distance of about fifty miles from Lake Marsh. Lake Labarge is thirty-one miles long, with an average width of five miles, and is very windy. It is the last of the remarkable series of lakes beginning with Lake Lindeman in the south. And here attention should be drawn to the singular conformation of the country which makes the springs no farther distant than thirty miles from tidewater on the south find their outlet in the great system of rivers which pour their waters through the Yukon into Bering Sea thousands of miles away.

The Hootalinqua River enters the Lewis twenty-eight miles below Lake Labarge and has acquired an interest apart from its size owing to the fact that it was the limit of the journey of the earliest prospector for gold in this region. Thirty-one miles farther down is the Big Salmon, and thirty-five miles still farther comes the Little Salmon, both of which are great streams for fishing, many Indians spending the summer months on the larger river preparing their winter salmon. After proceeding eighty miles farther the argonauts come to old Fort Selkirk, at the junction of the Pelly and Lewis Rivers, where there is a trading post. This is the farthest point to which the shallow boats which ply the Yukon reach, and the P. B. Weare, which will be a familiar name no doubt to those miners hereafter who endeavor to reach the gold fields by the water route, has been accustomed to winter. Ninety-six miles farther down the White River, which is described as the most wonderful of all the great system, enters the Yukon from the west. The volume of water is vast; it is muddy in color, and the current flows at the rate of eight or ten miles an hour. It discharges itself into the Yukon with such force that the roar can be heard for a long distance, and it muddies the larger river until the waters of the two can hardly be distinguished. The White River comes from a glacier region and is supposed to flow over volcanic deposits, but the meagreness of the information which exists in regard to this whole interior country appears in the fact that little more than has been said is known about one of the largest and most remarkable streams in the territory of the United States. Ten miles farther down the Yukon receives the waters of the Stewart River, along which rich finds of gold have recently been made. It is a quartz forma-

SEEKING THE POT OF GOLD. 57

tion and the rock assays $300. Seventy miles farther Sixty Mile Creek joins the swelling stream and fifty miles beyond Sixty Mile Creek the Klondike River enters from the east. The Yukon between the Klondike River on the east, and where Sixty Mile Creek enters it on the west, runs almost directly north and south. The gold discoveries on Sixty Mile Creek have been far to the west on the American side of the boundary, while the discoveries on the Klondike River have been to the east and altogether on Canadian soil. Continuing down the river from Koldnike the traveler would come to Forty Mile Creek, which a year ago was the centre of such gold mining excitement as there was, but for the present at any rate no seeker after wealth will venture a step beyond the Klondike region. The reports of miners coming from the gold fields all agree that the placer diggings along Forty Mile Creek, Sixty Mile Creek and Birch Creek have been abandoned for the more spectacular, sensational findings on the Klondike River. That Circle City is occupied by only a stray miner or tow, and that Forty Mile Post, which in 1895 boasted ten saloons, two restaurants, three billiard halls, two dance houses, an opera house, a cigar factory, a barber shop, two bakeries, several breweries and distilleries and a store, is now a deserted camp.

This desertion of Forty Mile Post and of Circle City, which is one hundred and seventy-five miles farther down the river, is believed by mining experts to be temporary, for the fields which feed them are practically exhaustless, although they have been abandoned now for diggings which will yield speedier returns.

But for the present the traveler may be safely left at Klondike, which was his original destination, having spent seven weeks in traversing the 650 miles between Lynn Canal and Dawson City, with dangerous and exciting experience, through swift and treacherous currents, log jams, floating ice and debris, whirl pools and rapids and dark canons full of unknown difficulties. The quickest time which can be made under existing conditions between Juneau and Dawson City is about a month.

Those who wish to take the route by way of St. Michael can board the steamer at San Francisco or Seattle, travel twenty-five hundred miles to St. Michael, which is the Alaskan seaport near the mouth of the Yukon River, then travel on the little river steamer 1895 miles clear across American territory and well into British Columbia. This trip takes about thirty days and the traveler is subject to tedious delays caused by ice jams and sand bars, so that by the time he reaches the gold field he is hardly in condition to take advantage of his opportunities. The period during which the Yukon River is navigable is so short that some think it hardly pays to attempt the journey in this way, although hundreds have essayed the trip in the first flush of the gold excitement. The ice does not break up at the mouth of the river earlier than the first of June and by the time the traveler reaches the fields and locates his claim winter is almost ready to set in and he is obliged to exist as best he can through the bitter cold of Arctic days. So it is that the majority of prospectors will continue to avail themselves of the overland trip from Juneau which has been described in detail.

A new route to the Klondike will be opened next spring. It is overland from Juneau to Fort Selkirk, on the Yukon, and is entirely by land. Captain Goodall, of the Pacific Coast Steamship Company, inspected it this summer and reported it practicable. It is about 700 miles long, and it crosses the divide over Chillkoot Pass, which is about fifty-five miles to the east of Chilkoot Pass. No lakes or rivers are on the route, but the trail runs over a high, level prairie. Old pioneer Dalton, after whom the trail is named, is now driving a band of sheep on the trail to Dawson City, where he expects to arrive in August with fresh meat for the miners. This Dalton trail is well dapted for driving stock, but for men to tramp it is believed to be too long.

One who is now at the Klondike diggings writes from there of his journey overland as follows:

"We arrived here from Dyea after seventy days of the hardest travel I ever experienced. We had all our provisions in cachet at Chilkoot Pass. We loaded everything on three sleds and turned them loose down the three-mile declivity. They landed all safe at the bottom on the Yukon side.

"Then we followed, winging and tumbling after. We crossed Lake Lindeman on the ice all right at the foot of the mountains and got safely to the head of Lake Bennett. By this time the weather was getting warmer and the snow melting. The snow crust on the lake would support the sleds, but we broke through at every step, and there was about a foot of slush under tne crust. After wading this way for two days and having traversed but four miles we went into camp to wait for a cold snap or more of a thaw to break up the ice. We lay in camp for three days, and then came a cold spell, the wind blowing a gale.

"When we struck Marsh Lake the weather had become warm again, and it took us three days to make seven miles through eight inches of slush, so we waded into a good patch of timber and remained there fourteen days building a boat. It took us six days to fell the trees and saw the boards out.

"When we got to the great Yukon we launched our little craft and tried her in the swift current of the mighty river (a river as large as the Mississippi) and found she would answer our purpose very well. The next day we came to a canon called 'Miller's Canon,' the most dangerous place on the river, where many a party have lost all they had, and their lives, too. It is a steep cut through the mountain range. The water rushes through with frightful speed. There is a long, devious way around the canon by land which requires four days' hard work to get over, while to shoot the canon only takes two and one-half minutes.

"As soon as the boat entered the canon she seemed to shiver and then plunged head foremost into the first waves, and about a half barrel of water came over the bow. Then she straightened out and rode through the rapids without shipping a drop more water. We continued down the river to Lake Labarge, thirty-five miles. There our boat riding ended for the present, the lake being still frozen solid. It is thirty miles long. The ice was smooth as glass, so we rigged up two sails on the boat (which we had deposited on two sleds).

"Two days later we once more launched into the friendly Yukon and floated calmly down the river to Klondike, a distance of four hundred miles from the last lake, in eight days."

There is talk already of building a railroad into the gold diggings, and the Canadian Government has been asked to help. An appropriation of $5000 was passed by the present Parliament to send surveyors into the field.

Two routes are suggested—one from a point on the Canadian Pacific, the other from Dyea.

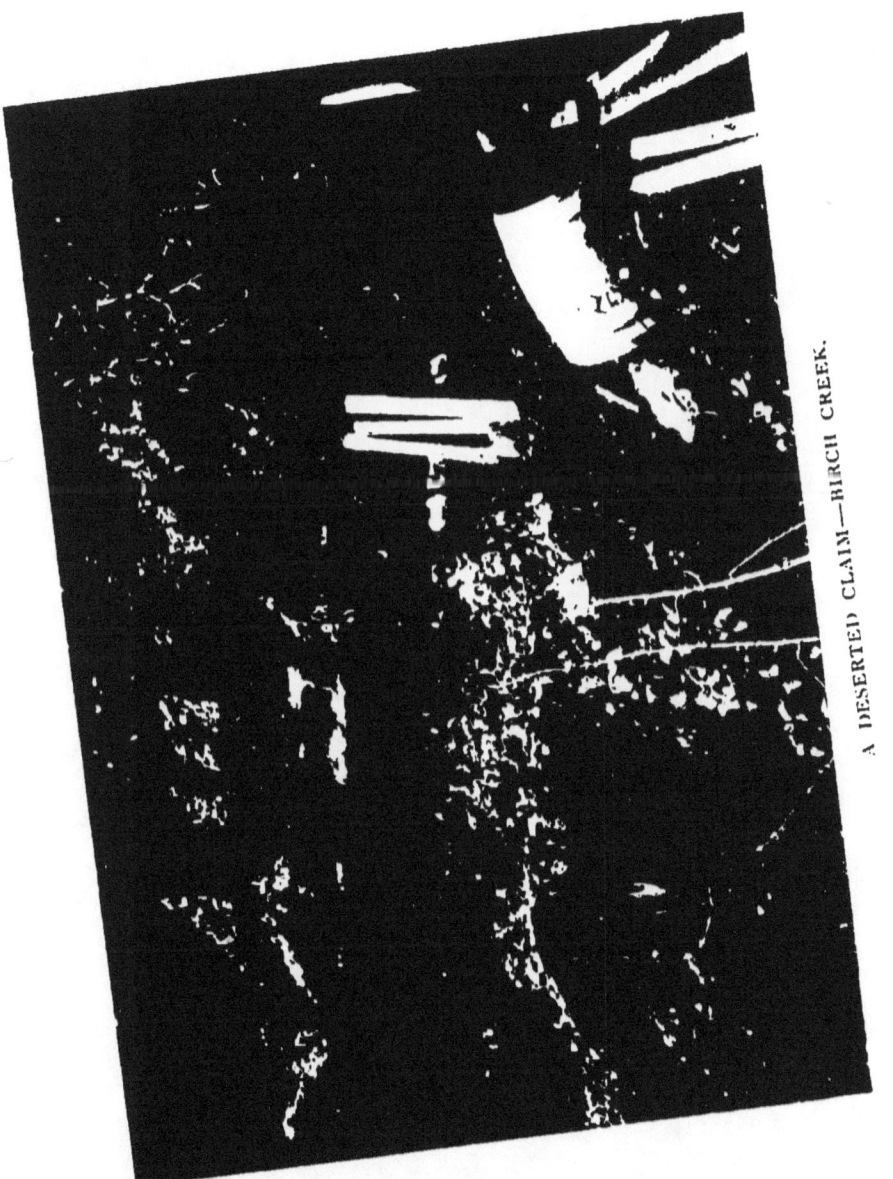
A DESERTED CLAIM—BIRCH CREEK.

It is said that neither offers serious difficulties from an engineering point of view. From Dyea only eighty miles of road would have to be built, the rest of the route being to the mines by means of the lakes and rivers. This road would abolish the peril of the Chilkoot Pass. The other route is 500 miles long and entirely within the jurisdiction of the Dominion of Canada, while the Dyea route would have its terminus in the soil of the United States. The day may not be far distant when the Alaskan country will be traversed by rail from the Canadian Pacific to Bering Straits. American enterprise may run a road all along the coast from Seattle to Asia.

CHAPTER IV.

LIFE IN CAMP.

A mining camp is always a spot of intense human interest. It is the breaking of the frontier —the first contact of civilization with the wilderness—and it brings into play all the rough elemental qualities of the human animal. The Yukon mining camps have been little worlds by themselves, isolated and ice-bound, and they have been rich in incident, though from all accounts they seem to have lacked the easy indifference to the sanctity of human life which characterized the earlier mining camps of California and Colorado. Forty Mile Post, for example, has been described as a characteristic gold town in every way but one. It boasts the company stores, an opera house, a barber shop, two bakeries, two restaurants, three billiard parlors, two terpsichorean resorts, several distilleries and ten saloons. Its exceptional feature is the utter absence of that lawlessness and disorder always looked for in frontier places. This same peaceful state of affairs obtains throughout the country. Law there is none, except miner's law, that stern, Draconian code, which decrees the extreme penalty for the least offense. The fact that there has never been a lynching or shooting affray there is testimony of the efficiency of self-government, where the consent of the governed has been secured.

It is not unlikely that some part of the general obedience is due to the liberality with which the moral obligation is construed. The Yukon Decalogue contains rather less than ten commandments. Thou shalt not avoid thy just debts: thou shalt not kill; thou shalt not steal; thou shalt not covet thy neighbor's claim, nor his sluiceboxes, nor his cabin, nor his mission squaw, nor anything that is his, make up the prohibited list.

One can hang a sack of gold dust outside of his cabin and it is perfectly safe. One saloonkeeper has $160,000 in gold in a little shack and he never locks his door.

A returning traveler says the only reminder of law and vested authority that he saw on the entire journey down the Yukon was at Forty Mile, or, to speak more precisely, at Fort Cudahy, whcih is across the bend of the river a mile or two away from the former place. There was a low stockade and a flagpole with the union jack flying.

There is a detachment of twenty-five Canadian mounted police stationed here and a magistrate, and the whole machinery of the law as applied to territories is in operation. They have very little to do in maintining order, and the police may be pardoned for assuming a little commission on the side, as it were, in going over the line into American territory to put Messres. Van Wagenen and Hestwood in possession of their mine, which was held by indignant miners.

The police are a well-equiped and well-drilled body of men, armed with Lee-Metford rifles. As cavalry or mounted police they are out of their element, as it is impracticable to use horses here.

It is now proposed by the United States Government to establish an army post in the neighborhood of the diggings, with headquarters probably at Circle City. The troops will act as police.

There is a marked difference between the attitude of the two governments toward their pioneers. Four-fifths of the men in the interior are Americans, and more than two-thirds of the whole number have been in American territory. On the British side, with one-fourth the interest at stake, the Canadians have a picked and athletic body of men ready to respond in any emergency. Should disaster befall any man or body of men within the Dominion's jurisdiction, these police would hasten to the rescue as rapidly as it is within human power to do, and without any question as to whether or not the unfortunates were citizens.

Over the line, in Alaska, is a stretch of country where two or three New Englands might be

LIFE IN CAMP.

thrown in at random without touching, the dignity of the United States is upheld by one man, a customs officer, whose duties partake of those of a tax collector and a detective combined. The only solicitude expressed is in the way of collecting taxes. Recently a United States postmaster has been added to the official life, but naturally he has nothing to do except handle mail. A United States commissioner is the latest promised acquisition, although he has not yet put in a formal appearance.

The most prevalent trouble is scurvy, which results from scarcity of vegetables and fresh meats. A diet of beans, salt pork and bad bacon brings trouble. Fresh meat is always scarce. Moose and caribou have been killed off and the chase would not supply a fraction of the population. There are graylings and other fish in the Yukon and they can be hooked through the ice, but few will stand out in the middle of a river at 60 degrees below zero and with time worth $15 a day. Last winter a quarter of beef was sledded into Circle City with dogs. It was viewed with wonder at the store for a while and then raffled off for $400 for the benefit of a projected miners' hospital. This spring an enterprising

Juneau man drove forty head of cattle in from the coast—800 miles—and beef went at 50 and then at 70 cents a pound. If anybody gets sick there are patent medicines in the stores, and four or five doctors who diagnose a patient's claim before presenting the bill.

Winter in the Yukon Basin is not altogether an unbearable season. The thermometer often falls to 70 and even 80 degrees below zero, but there is neither wind nor moisture, and the extreme cold is not then realized. When working out of doors a miner wears a thermometer as he wears a watch. He consults it every now and then for prudence's sake, and when the mercury freezes he knows that it is time to go in.

Most miners adopt the native dress of skin trousers and parka. The best of these shirt-like garments are brought over from Siberia, and find ready sale at $25.

There are two kinds of boots, the water boot, made of seal and walrus skins, and the dry weather or winter boot, made in all sorts of fashions, some with picturesque fur trimming. The boots as a rule are the handiwork of the coast Indians. They range in price from $2 to $5 a pair. Trousers are made of Siberian fawn skin and the skin

LIFE IN CAMP.

of the marmot or ground squirrel. The parka or upper garment is usually made of marmot skins and trimmed with wolverine around the hood and lower edge. These parkas are sometimes very elaborate, with hair six inches in length hanging from the hood to protect the face, or made of fawn skins and trimmed with the fur of the white wolf. These elaborate parkas are usually worn by the women and differ in shape somewhat from those worn by the men. They are sometimes beautifully embroidered with colored skins and ornamented with otter's fur and dyed feathers, and they have been known to cost as high as $100 apiece. Flannels are worn underneath and the dress is described by those who have worn it as weighing less than the ordinary clothes of a country where the thermometer only falls to zero.

Women who have drifted in from the coast received an odd rebuke from Captain Constantine, of the Territorial police. The women naturally put on bloomers in coming over the mountains, and when they got on the other side they continued to wear bloomers altogether. Bloomers were more than Captain Constantine would stand, and he gave orders that if the bloomers did not go the wearers would.

Help is scarce. Indians who cannot speak either English or Chinook receive $60 per month and all the tobacco they can use. These are willing to help, but with the judgment of children. Every white man that will act as boss of a gang is pressed into service.

Gold dust and nuggets take the place of currency in the new diggings and throughout the Yukon Basin. There is little money in circulation. Every man carries a pair of gold scales, and people learn to make change as quickly as with coin. A hair cut costs 75 cents in gold dust, a glass of whisky 50 cents, and during the winter season, when the thermometer ranges between zero and 70 degrees below, whisky is sometimes sold in solid blocks. The established value of gold dust is $17 an ounce. Nuggets of one and two ounces are by no means uncommon.

The principal sport with mining men is found around the gambling table. There they gather after nightfall and play until the late hours in the morning. They have some big games, too. It sometimes costs as much as fifty dollars to draw a card. A game with $2000 as stakes is an ordinary event. But with all that there has not been any decided trouble. If a man is fussy

LIFE IN CAMP.

and quarrelsome, he is quietly told to get out of the game, and that is the end of it.

Drinks are 50 cents, and returned miners say that when they left some of the saloons were taking in $1000 to $2000 a day.

Whisky will be plentiful hereafter, even if food is not. One trader has secured a permit to send in 2000 gallons of liquor.

The Alaska Commercial Company and the Northwestern Transportation and Trading Company have each received permission to ship across the border 5000 gallons.

Many people have an idea that Dawson City is completely isolated and can communicate with the outside world only once in twelve months. That is a mistake. Circle City, only a few miles away, has a mail once each month, and there the Dawson City men have their mail addressed. It is true the cost is pretty high, one dollar a letter and two dollars for a paper; yet by that expenditure of money they are able to keep in direct communication with their friends on the outside.

The camp is at present without any public institutions, but by next season they will have a church, a music hall, a school house and a hospital. This last institution will be under the direct control of the Sisters of Mercy, who have already been stationed for a long time at Circle City and Forty Mile Camp. Nearly a score of children were in Dawson City when the last party left, and Joseph Ladue, who owns the town site, donated a lot and one hundred dollars for a school. No one can buy anything on credit in Dawson. It is spot cash for every one, and payment is always gold dust. Very few have any regular money.

The mosquito is an almost intolerable pest. In the Yukon region he is so small that the finest netting cannot keep him out, but his voracity is seemingly boundless.

During the summer this pest gives the population no rest. The deepest canon and the loftiest mountain top, the open ground or the thickest forest being equally infested. The only relief, if it can be called relief, is when the winds blow the insects to less windy altitudes; but it is not an every day occurrence for the wind to blow. Lieutenant Schwatka, in his account of his trip to Alaska, says that bears under stress of hunger sometimes come down to the river in mosquito season and are attacked by swarms of insects, which sting them about the eyes so that they go blind and die of starvation.

LIFE IN CAMP.

There is one side of the Klondike picture which has been kept in the background, but about which whispers are beginning to be heard. It is a picture of suffering and starvation. One of the returned fortune makers is quoted as saying:

"You would find it easier to believe the most wonderful yarns I could tell you of the wealth of the country than some of the hardships I have known many men to undergo. Men can suffer a great deal and almost forget it if they eventually become rich, but for every man who has returned with a sack of dust there are now one hundred poor devils stranded and starving in that country.

"When I say starving I mean it literally. It seems incredible that a man would see another —his neighbor, at that—slowly dying by inches for want of food and deliberately refuse him a pound of bacon or pint of beans, yet that thing is happening every day, and God only knows how many frozen corpses will make food for wolves on Klondike this winter. When I left there was not enough food in the country to supply those already there, and as boats cannot take in much more before the river freezes, how are hundreds now on their way there to exist? It is not that men are selfish or avaricious, but few of the old miners have more than enough to keep them through the winter, and it is only a question of preserving their own lives or those of others."

It is likely to be as bad next winter. The united efforts of the Alaska Commercial Company and the North American Transportation and Trading Company cannot transport over 4500 tons of freight up the river this season, and not until next February can stuff be freighted over from Dyea, Juneau and other points down along the southern coast. Prices for food and other supplies were almost beyond belief last winter. Flour was $120 a hundred weight at one time, and beef from $1 to $2 a pound. Moose hams sold for about $30, or $2 a pound. Ordinary shovels for digging brought $17 and $18 apiece. A few crates of eggs were brought in about March 1 by pack horses, and these sold readily for $3 to $5 a dozen. They were not fresh by any means.

Wages, however, were proportional; $2 per hour were common wages and even in the summer a man can command $1.50 per hour, or from $15 to $20 a day.

LIFE IN CAMP.

A new arrival at Dawson City, writing to his brother, says:

"This is a great camp, and a conservative estimate of its richness sounds like exaggeration. I have been here now twelve days and cannot get a hold of anything. I cannot even buy a foot of ground in the town, not to mention the diggings, values are so extremely high. Every foot of ground in this district is claimed, and there are hundreds of prospectors in the adjacent country looking for other rich ground. The gravel must be very rich in gold or nobody wants it. From the amount of gold dust and nuggets I have seen in Klondike, and the mad hunt for it, the district must be all they claim for it."

The mines of the Yukon are of a class by themselves, and it is necessary to follow new methods for getting the gold. To begin with, the ground is frozen. From the roots of the moss, which is often a foot thick, to the greatest depth that ever has been reached the ground is as hard as a bone. The gold is found in a certain drift of gravel, which lies at varying depths, often as far down as twenty feet. Only that portion of the gravel just above hard pan—by which is usually meant clay—carries gold in any quantity, and in favored localities this particular gravel is extraordinarily rich.

As in nearly all placer mines, the low places of what has formerly been the bed of the creek are the richest, the deposits decreasing toward the outer edges.

The size of a claim is fixed by agreement among the miners of any particular locality. It is a section of the creek of a certain length—sometimes 200 feet, sometimes 500—and it extends from rim to rim in width. The reason of this variableness in the size of the claims on the different creeks is that on some a greater length is required to make them worth a man's while to work them. The paying deposits may be scattered so a man could make wages only by working here and there over a large territory. Of course, the conditions surrounding the first discovery made on a creek are the basis for fixing the size of a claim on that stream. The discoverer of a new field is allowed two claims, while all others are permitted to take but one at a time. However, when a locater has worked out his assessment of a few days' work he is at liberty to take another. When a sufficient number of men

LIFE IN CAMP.

arrive on a new creek to make it impracticable to work together in harmony without organization, they hold a meeting and elect one of their number as a register or clerk, and thereafter a record is made of all locations and all transfers, for which a small fee is charged.

In prospecting the usual method is followed, i. e., sinking holes to bed rock across the stream and testing the dirt until the pay streak is found.

Having located his claim, the miner scrapes off as much moss as he can, and, turning a stream of water on to the frozen ground, gradually thaws, scrapes and digs his ditch. The gold lies at bed rock, fifteen to twenty feet below the surface. A drainage ditch must then be dug, a dam built and sluice boxes placed.

Winter mining has been experimented with to some extent. Work cannot be started until the cold weather is settled beyond the possibility of a surface thaw, nor can it be continued beyond the first promise of spring. A fire is built and kept burning until the ground beneath is thawed to bed rock, after which the drift is removed, leaving a hole several feet wide. By banking the fires against the side of the hole every night and removing the soft earth next morning, a tunnel is formed. A foot and a half a day is as much as the greatest industry can accomplish, but that amounts to 150 feet in the season. The pay dirt is piled up and is not washed until the following spring.

CHAPTER V.

MINING EXPERTS AND SCIENTISTS.

Professor N. S. Shaler, who is perhaps the best living American authority on geology, has been telling his classes at Harvard for the last twenty years that the coming great discoveries of gold on this continent would be in Alaska. The possibilities for bonanza finds among the Sierras, he explained, had been narrowed to a point where there was little opportunity except to develop known veins, but in the great extension of the Rocky Mountain system to the North there doubtless lay the mother vein, which sooner or later would come to light.

Professor Shaler's prophecy, based on scientific deductions, has come true, and other scientists now agree with him that the Alaskan country contains limitless possibilities for the discovery of gold.

And not the scientists alone. So hard-headed a pioneer as John W. Mackay, the last and greatest of the bonanza kings, who went into the California gold fields and dug out a fabulous fortune, which has been growing ever since, expresses his belief in the reports of the marvelous richness of the newly-discovered fields.

"I have no reason to doubt them," he says. "I have had great confidence in the mining possibilities in British Columbia and Alaska—have always believed that those frozen, almost inaccessible regions contain heavy deposits of precious metals. Some enormous 'finds' of gold have undoubtedly been made there, and yet we know little or nothing of the possibilities of the country. Think of Williams' Creek, for instance, in the Caribou region in British Columbia. As long ago as 1860 something like fifty millions of gold were taken out. It was placer mining there, just the same as the Klondike."

Mr. Mackay believes that in time modern mining methods will be carried up into the Yukon country, and that all parts of the country will be opened. "Capital," he says, "will always go where there is a chance for legitimate investment, and transportation facilities will increase as rapidly as the travelers."

Mr. Mackay thinks the excitement over the discoveries may increase. "I see in it," he says, "something like the excitement of the early fifties over the gold discoveries of the Pacific coast region. The reports of rich individual finds are likely to continue, and the arrival of every ship loaded with fortunate gold hunters will stimulate the imagination, hopes and desires of the would-be gold hunters. We hear nothing of the failures. One man who is lucky is more talked about than a thousand who fail."

Mr. Mackay says that his experience in California was that about one man in ten used to get on, and by "getting on" he means not becoming a millionaire, but making a living and a little more.

R. E. Preston, the Director of the United States Mint, has become convinced of the great possibilities in the Klondike region. While he thinks it is as yet too early to hail the Klondike as a new Eldorado, he says the history of gold production in Alaska hitherto would prepare the mind for the acceptance of a belief in the likelihood of further gold discoveries in that region or its proximity.

"The gold product of Alaska thus far," he says, "has been remarkable rather for its regularity than its amount, and is therefore more favorable to the permanency of development of the mineral resources than if it were subject to violent fluctuation.

"Nature seems to have sprinkled Alaska and all Asiatic Russia with gold. The latter region sends annually over $25,000,000 to the mint at St. Petersburg. The production of gold there is such that the annual output of the Russian Empire would, it is claimed, exceed $50,000,000 were it not for the obstacles put in the way of human industry by an inclement climate and an inhospitable soil."

Dr. W. H. Dall, of the Smithsonian Institution at Washington, who has for years been regarded as the highest authority on the Alaskan country and who is a geologist of note, says he has no doubt of the truth of the stories told of the richness of the Yukon soil.

"The gold-bearing belt of Northwestern America," he says, "contains all the gold fields extending into British Columbia and what is known as the Northwest Territories and Alaska. The Yukon really runs along in that belt for 500 or 600 miles. The bed of the main river is in the valley. The yellow metal is not found in paying quantities in the main river, but in the small streams which cut through the mountains on either side. Mud and mineral matter are carried into the main river, while the gold is left on the rough bottom of these side streams. In most cases the gold lies at the bottom of thick gravel deposits. The gold is covered with frozen gravel in the winter. During the summer until the snow is all melted, the surface is covered with muddy torrents. When summer is over and the springs begin to freeze, the streams dry up. At the approach of winter, in order to get at the gold the miners find it necessary to dig into the gravel formation."

George Frederick Wright, professor of geology at Oberlin College, thinks that the "mother lode" may be looked for successfully in Alaska. In his opinion it exists somewhere up the streams on which the placer mines are found. The source of the Klondike gold, he says, is from the south, and the gold was doubtless

FISHING VILLAGE ON THE YUKON.

transported by glacier action. The Klondike region is on the north side of the St. Elias Alps, and the glaciers flowed both north and south from these summits.

"Placer mines," says Professor Wright, "originate in the disintegration of gold-bearing quartz veins, or mass like that at Juneau. Under subaerial agencies these become dissolved. Then the glaciers transport the material as far as they go, when the floods of water carry it on still further. Gold, being heavier than the other materials associated with it, lodges in the crevices or in the rough places at the bottom of the streams. So to speak, nature has stamped and 'panned' the gravel first and prepared the way for man to finish the work. The amount of gold found in the placer mines is evidence not so much, perhaps, of a very rich vein as of the disintegration of a very large vein."

"What the prospectors have found points to more. The unexplored region is immense. The mountains to the south are young, having been elevated very much since the climax of the glacial period. With these discoveries and the success in introducing reindeer, Alaska bids fair to support a population eventually of several millions."

William Van Slooten, an eminent mining engineer and metallurgist, sees in the reports from the Klondike indications of a more extraordinary deposit of gold than that of California. He says:

"No such specifically large amounts of gold were taken out by individuals during any similar period of California gold hunting. Two months of work in the water has realized more than any six months heretofore known in the history of gold mining.

"We had long been aware that there was gold in the Yukon basin, but the total output for the last ten years before the Klondike developments amounted to not more than a million dollars' worth at the utmost. Now, within two months, five millions have been taken out of the Klondike regions. It took the first eight months of work in California to pan out that amount under infinitely more favorable conditions of climate and weather. That is a straw worth noting."

The latest and therefore the most important official investigation of the gold fields is that conducted under the auspices of the United States Geological Survey in 1896 by J. Edward Spurr, accompanied by two geologic assistants.

The expedition was sent out in accordance with an appropriation by Congress of $5000 for the investigation of the coal and gold resources of Alaska. A like appropriation for the year before resulted in the expedition headed by Dr. George F. Becker, which investigated the gold fields of Southern Alaska. Mr. Spurr's party crossed the Chilkoot Pass about the middle of June and passed down the Yukon in a small, roughly-built boat to the crossing of Forty-Mile Creek. A summary of his report was submitted to Congress by the director of the Geological Survey through the Secretary of the Interior February 2, 1897. Mr. Spurr's party and Dr. Becker's both took numerous photographs along the routes they traversed. It appears from Mr. Spurr's report that the gold belt is likely to be found running in a direction a little west of northwest.

Running in a direction a little west of northwest through the territory examined is a broad, continuous belt of highly altered rocks. To the east this belt is known to be continuous for 100 miles or more in British territory. The rocks constituting this belt are mostly crystalline schists associated with marbles and sheared quartzites, indicating a sedimentary origin for a large part of the series. In the upper part a few plant remains were found, which suggest that this portion is probably of Devonian age. These altered sedimentary rocks have been shattered by volcanic action, and they are pierced by many dikes of eruptive rock. Besides the minor volcanic disturbances, there have been others on a large scale, which have resulted in the formation of continuous ridges or mountain ranges. In this process of mountain building the sedimentary rocks have been subjected to such pressure and to such alteration from attendant forces that they have been squeezed into the condition of schist, and often partly or wholly crystallized, so that their original character has in some cases entirely disappeared. In summarizing, it may be said that the rocks of the gold belt of Alaska consist largely of sedimentary beds older than the Carboniferous period; that these beds have undergone extensive alteration, and have been elevated into mountain ranges and cut through by a variety of igneous rocks.

Throughout these altered rocks there are found veins of quartz often carrying pyrite and gold. It appears that these quartz veins were formed during the disturbance attending the uplift and alteration of the beds. Many of the veins have been cut, sheared and torn into fragments by the force that has transformed the sedimentary rocks into crystalline schist; but there are others, containing gold, silver and copper, that have not been very much disturbed or broken. These more continuous ore-bearing zones have not the character of ordinary quartz veins, although they contain much silica. Instead of the usual white quartz veins, the ore occurs in a sheared and altered zone of rock and gradually runs out on both sides. So far as yet known, these continuous zones of ore are of relatively low grade. Concerning the veins of white quartz first mentioned, it is certain that most of them which contain gold carry it only in small quantity, and yet some few are known to be very rich in places, and it is extremely probable that there are many in which the whole of the ore is of comparatively high grade.

No quartz or vein mining of any kind has yet been attempted in the Yukon district, mainly on account of the difficulty with which supplies, machinery and labor can be obtained; yet it is certain that there is a vast quantity of gold in these rocks, much of which could be profitably extracted under favorable conditions. The general character of the rocks and of the ore deposits is extremely like that of the gold-bearing formations along the southern coast of Alaska, in which the Treadwell and other mines are situated, and it is probable that the richness of the Yukon ocks is approximately equal to that of the coast belt. It may be added that the resources of the coast belt have been only partially explored.

Besides the gold found in the rocks of the Yukon district there is reason to expect paying quantities of other minerals. Deposits of silver-bearing lead have been found in a number of localities, and copper is also a constituent of many of the ores.

Since the formation of the veins and other deposits of the rocks of the gold belt an enormous length of time has elapsed. During that time the forces of erosion have stripped off the overlying rocks and exposed the metalliferous veins at the surface for long periods, and the rocks of the gold belt, with the veins which they include, have crumbled and been carried away by the streams, to be deposited in widely different places as gravels, or sands, or muds. As gold is the heaviest of all materials found in rock, it is concentrated in detritus which has been worked over by stream action; and the richness of the placers depends upon the available gold supply, the amount of available detritus, and the character of the streams which caryy this detritus away. In Alaska the streams have been carrying away the gold from the metalliferous belt for a very long period, so that particles of the precious metal are found in nearly all parts of the Territory. It is only in the immediate vicinity of the gold-bearing belt, however, that the particles of gold are large and plentiful enough to repay working, under present conditions. Where a stream heads in the gold belt, the richest diggings are likely to be near its extreme upper part.

In this upper part the current is so swift that the lighter material and the finer gold are carried away, leaving in many places a rich deposit of coarse gold overlain by coarse gravel, the pebbles being so large as to hinder rapid transportation by water. It is under such conditions that the diggings which are now being worked are found, with some unimportant exceptions. The rich gulches of the Forty Mile district and of the Birch Creek district, as well as other fields of less importance, all head in the gold-bearing formation.

A short distance below the heads of these gulches the stream valley broadens and the gravels contain finer gold more widely distributed. Along certain parts of the stream this finer gold is concentrated by favorable currents and is often profitably washed, this kind of deposit coming under the head of "bar diggings." The gold in these more extensive gravels is often present in sufficient quantity to encourage the hope of successful extraction at some future time, when the work can be done more cheaply and with suitable machinery. The extent of these gravels which are of possible value is very great. As the field of observation is extended farther and farther from the gold-bearing belt, the gold occurs in finer and finer condition, until it is found only in extremely small flakes, so light that they can be carried long distances by the current.

It may be stated, therefore, as a general rule, that the profitable gravels are found in the vicinity of the gold bearing rock.

The gold-bearing belt forms a range of low

mountains, and on the flanks of these mountains, to the northeast and to the southwest, lie various younger rocks which range in age from Carboniferous to very recent Tertiary, and are made up mostly of conglomerates, sandstones and shales, with some volcanic material. These rocks were formed subsequent to the ore deposition, and therefore do not contain metalliferous veins. They have been partly derived, however, from detritus worn from the gold-bearing belt during the long period that it has been exposed to erosion, and some of them contain gold derived from the more ancient rocks and concentrated in the same way as is the gold in the present river gravels. In one or two places it is certain that these conglomerates are really fossil placers, and this source of supply may eventually turn out to the very important.

In the younger rocks which overlie the gold-bearing series there are beds of black, hard, glossy, very pure lignitic coal. An area of these coal-bearing strata lies very close to the gold-bearing district, in the northern part of the region examined, and as the beds of coal are often of considerable thickness and the coal in some of them leaves very little ash and contains volatile constituents in considerable amount, it is probable that the coal deposits will become an important factor in the development of the country.

There were probably 2000 miners in the Yukon district during the past season, the larger number of whom were actually engaged in washing gold. Probably 1500 of them were working in American territory, although the migration from one district to another is so rapid that one year the larger part of the population may be in American territory and the next year in British. As a rule, however, the miners prefer the American side, on account of the difference in mining laws. These miners, with few exceptions, were engaged in gulch digging. The high price of provisions and other necessaries raises the price of ordinary labor in the mines to $10 per day, and therefore no mine which pays less than this to each man working can be even temporarily handled. Yet in spite of these difficulties there were probably taken out of the Yukon district the past season, mostly from American territory, approximately $1,000,000 worth of gold.

An overland route should be surveyed and constructed to the interior of Alaska. All the best routes which can be suggested pass through

British territory, and the co-operation of the two governments would be mutually beneficial, since the gold belt lies partly in American and partly in British possessions. At the present time Mr. Spurr thinks that the best route lies from Juneau by way of the Chilkat Pass overland to the Yukon at the junction with the Pelly. This trail is the Dalton trail which has already been described, and it is said to open up a good grazing country and no great obstacles to overcome. The Chilkat Pass is considerably lower than the Chilkoot, over which the Geological Survey party of 1896 passed. If a wagon road, or even a good horse trail, could be built as indicated, the cost of provisions and other supplies would be greatly reduced, many gravels now useless could be profitably worked, and employment would be afforded for many men. With the greater development of placer diggings would come the development of mines in the bed rock.

Besides the coal which has been alluded to there is abundant timber throughout the whole of the interior of Alaska, along the valleys of the Yukon. For four or five months in the summer the climate is hardly to be distinguished from that of the northern United States—Minnesota or Montana, for example, and although the winters are very severe, the snowfall is not heavy. Work could be carried on underground throughout the whole of the year quite as well as in the higher mountains of Colorado.

The area hastily examined during the past season is but a portion of the great interior of Alaska. That gold occurs over a large extent of country has been determined, but the richness of the various veins and lodes remains to be ascertained by actual mining operations. Gold is known to occur in the great unexplored regions south of the Yukon, because of its presence in the wash of the streams, and it is quite probable that the Yukon gold belt extends to the north and west; but this can be determined only by further exploration.

CHAPTER VI.

PLACER MINING AND HYDRAULICS.

There are four stages in the development of newly-discovered gold fields, such as those which have been brought to light in the Yukon Basin.

First come the men with crude outfits and few

PLACER MINING AND HYDRAULICS.   95

resources, who, with pan and pick, gather the gold that lies near the surface, washing out the grosser earths and leaving the precious metal by itself. This is placer mining in its simple form.

After the gold lying on the surface and most readily at hand has been exhausted a little more complicated process is called into play. This is conducted by groups or associations of miners who use "long Toms" and cradles.

Hydraulic mining is the third stage. In hydraulics water is brought from a long distance and applied to the pay dirt at great pressure in order to separate the gold from the dross.

Last of all comes quartz mining, or tearing the gold by main force out of its beds in the rock beneath and separating it by means of stamps and pestles.

In the Yukon region the process has not yet passed the first stage, and so rich are the finds there and so difficult the importation of machinery and supplies that it may be years before the last stages will become available, although the never-satiated thirst for gold, combined with modern enterprise and ingenuity, is likely to make even the frozen rocks of Alaska amenable to modern appliances.

The history of placer mining is full of romance. It is as old as the world itself, if any reliance can be placed upon the traditions that have come down to us from prehistoric times. Gold dust and nuggets came in exchange to the Greeks from the barbarians of the north centuries before the birth of Christ, and it has been surmised that the precious metal was taken out of the mines in Siberia and in the Ural Mountains, which still yield so generously. The first placer mining of which there is any record was carried on by digging the sand or gravel, mixing it thoroughly with water, and then pouring it over floating platforms covered with skins, in which the gold settled, while the lighter sand flowed off with the water. To this practice we doubtless owe the mythological story of the journey of Jason with his Argonauts in search of the Golden Fleece. The Golden Fleece, it has been surmised, was simply the skin of the sheep which was used to catch these golden products of the placer miners. And it is significant that the voyage of the Argonauts was up the Black Sea or the Euxine into the very region of the Ural Mountain gold fields which have already been mentioned.

PLACER MINING AND HYDRAULICS.

In ancient times all gold was obtained by washing, and it has been only within recent years that the more difficult process of digging and smelting gold-bearing quartz has been resorted to. The wealth of the Indies consisted in golden sand, which their rivers washed down from the gold-bearing mountains. So it was with Russia, Africa, Australia and California. All the earlier mining, of which the records are so many and so fascinating, was done by placers in the old primitive manner. This was true especially of California. Mr. Preston, the Director of the United States Mint, estimates that 75 per cent. of the gold production of the United States between 1849 and 1865 was the result of placer mining. This would make a total of nearly $700,000,000 for the United States alone, to say nothing of the placers who are still at work in ever-diminishing numbers as the ore becomes more difficult to find. Ore is still being washed out in almost all the gold districts. California, Russia and Alaska are examples in point. There is even a little placer mining in Colorado, which has been distinctively the home of quartz mining from the beginning. Mr. Preston estimates that between fifteen and twenty per cent. of the Californian product is still the result of placer mining, and gives other percentages as follows:

Oregon, Washington, Montana and Idaho, 12 per cent.; Utah, 8 per cent.; New Mexico, 6 per cent.; Colorado, 1 per cent.

The South African mines are almost entirely quartz deposits.

The beginning of placer mining in America may be said to date from the discovery by James W. Marshall of pieces of gold while digging a race for a saw mill at Coloma, California, January 19, 1848. The announcement of his discovery was the signal for an influx or argonauts, and those who first landed in California had for implements only the pick, shovel, rocker and wheelbarrow. This is about the outfit of a miner in the Klondike region to-day. It was only a few months, however, before the necessities of the case compelled the introduction of what is known as the "Long Tom." This is a rough trough ten or twelve feet in length, narrow at the top and wide at the lower end, set on an incline, with an iron plate on the bottom perforated so that the gold will drop through as it is washed along. The "Long Tom" is really a development of the rocker or cradle. The rocker is

PLACER MINING ON MILLER'S CREEK.

what its name implies. It has a hopper at one end, with a perforated bottom, and this stands over an inclined canvas stretcher. The gravel is thrown into the hopper, water is poured over it and the cradle is rocked. In this way the fine sand and the gold fall through the holes on to the canvas; the gold sticks fast and the sand rolls away. The most primitive of all placer mining is the use of the pan, which consists simply in filling an ordinary pan with pay dirt, stirring it about very slowly and carefully, pouring water over the gravel at the same time, so as to wash away the lighter dirt and let the heavier gold sink to the bottom. The process is exceedingly slow, but in a region like the Klondike it is so full of striking possibilities as to be fascinating. One of those who have just returned from the Yukon describes how he found no less than a thousand dollars in gold dust at the bottom of one of these pans after washing away the dirt.

Placer mining, which depends so greatly upon the effect of water, would seem to be carried on under difficulties in the Yukon River Basin, where water is frozen solid during nearly ten months of the year, but the invention and industry of the Americans now on the field may be depended upon to bring even these hard conditions under their control, and it may be even that the miners there will be using hydraulic methods before very long.

Hydraulic mining is essentially the result of American inventive genius. It is the perfect development of the early form of placer mining as illustrated in the cradle and the rocker, for it may be said that the rocker, which is the rudest and simplest of all machines employed in the separation of gold from gravel, embodies all the essential features of the elaborate machinery used in hydraulic mining. For instance, the cradle is an oblong box, about four feet in length, mounted on a pair of transverse rockers and furnished with a set of graded sieves laid in tiers, "riffles," amalgamated plates and blankets, for the separation and arrest of the gold in its descent from the hopper into which the gold-bearing gravel is placed, to the outlet at the lower end. These devices are all present in hydraulic mining, but they are so enlarged as to be hardly recognizable. Hydraulic mining may be said to have had its origin in the invention of the flume by a Connecticut Yankee named Mattison in California three years after the discovery of gold. The flume was a very simple thing, consisting of a trough to bring water down the hillside from a ditch over where the mine was opened. The first flume gave the water a head of about forty feet, discharging it into a barrel, from the bottom of which depended a hose about six inches in diameter, made of common cowhide and ending in a tin tube about four feet long, which tapered to a point about an inch in diameter. With the head of water thus obtained, a stream turned dirt, washed off the lighter earth and gravel, while the coarser gravel was washed more carefully and thrown out with a sluice fork, the name of the stick used for that purpose. This flume was called a sluice. Later came the "ground sluice," which consisted in making the bed rock on which the pay dirt rested perform the duty of sluices, while a stream of water, used for washing away the dirt, was constantly trained against the bank. This water had about the same effect as water in any stream rubbing constantly and ceaselessly against its own banks where they offer resistance to the current.

It can be easily seen how modern hydraulic mining grew out of these comparatively simple contrivances. For the cowhide hose, canvas and then iron were substituted, and improvements have been constantly going on, until now it is estimated that $100,000,000 is invested in ditches, dams and tunnels in California alone. Water has been carried from almost incredible distances around apparently insurmountable obstacles so as to be brought into play for the washing of gold out of the gravel of arid diggings. In some instances from 250 to 300 miles of ditches and canals have been built at a cost of millions of dollars before water could be brought to play upon the gold-bearing dirt. Indeed it is an axiom among miners that the richness of the gravel is not so important as the abundance of water, for with water in sufficient quantities gravel containing even insignificant percentages of gold can be made to pay, and through the application of American inventiveness it has been found possible to wash out the deep gravel deposits on the high banks of the canons of streams where gold has been found. The beginning of this complete method of hydraulic mining is usually given as 1856. It was not until more than ten years after this that hydraulic mining was revolutionized by the introduction of the "monitor" in place of the discharge pipe of earlier days. After iron began

PLACER MINING AND HYDRAULICS. 103

to be employed for the flumes the pipes were gradually enlarged and strengthened, until they measure now from fifteen to thirty inches in diameter, terminating in monitors, which discharge the streams of water against the rocks with such tremendous force as to toss about like pebbles rocks which are tons in weight. The hydraulic monitor in action resembles very much a piece of military or naval ordnance. It is united to the supply pipe at the breech with a water-tight socket joint, which enables the miner to direct the nozzle toward any point. In spite of the tremendous force which the hydraulic monitor represents it can be managed almost by a child through a simple and effective arrangement called the "deflector." The deflector consists of a sleeve of sheet iron working on an elbow joint over the nozzle. To this sleeve is riveted an iron handle four or five feet long, by means of which the deflector can be moved so that the lip shall impinge on a column of water emerging from the nozzle of the monitor. An almost imperceptible angle is thus formed in a column of water which slowly moves the monitor in the opposite direction, relieves the friction and straightens the line of discharge. With all this tremendous force at work it is remarkable that modern hydraulic mining should have been carried to such a point of perfection that the amount of gold lost in washing is hardly worth taking account of, although in the old methods of placer mining it was estimated that from one-third to one-half of the fine gold was carried away in the debris. To illustrate the tremendous force of the water brought to bear upon the gold deposits through the hydraulic engines a correspondence, which was begun some years ago by Mr. Justice Field, of the Supreme Court, is of great interest. Justice Field's letter follows:

Washington, D. C., January 23, 1891.
Hon. James G. Fair:

Dear Sir:—Last evening I dined at General Schofield's and met the President (Harrison). There were a number of distinguished people present besides the President, among whom were the Chief Justice, the Speaker of the House of Representatives (Mr. Reed), Senators Sherman, Stanford and McMillan, Secretary of the Treasury Windom and Mr. McKinley and Mr. Wheeler of the House. During the evening the conversation turned upon California and her wonderful products and mining operations. I took occasion to speak of hydraulic mining and the wonderful manner in which the hills were torn down by hydraulic machinery. I stated that I had understood you to say that such was the force of the water thrown through a hose when it came from one hundred to two hundred feet in height that boulders weighing half a ton could be moved by streams playing upon them and that the force was sometimes so great that it would be impossible to cut the stream. At this statement much surprise was manifested, and I thought that a smile of incredulity passed over the features of the guests. Seeing this, I said that I would prove the facts stated in a communication to them.

Now I write to you for the information desired. Please send me some carefully prepared statistics as to hydraulic mining, particularly as to the power exerted of a column of water thrown by such machinery, and as to how large boulders can be moved by the force of the stream and on the point whether it is true that the force of the stream is sometimes so great that it cannot be cut. I would be much obliged if you could give me full particulars in regard to these matters in a communication that I can use if necessary. I propose to send a letter to each one of the guests, stating the facts, and thus remove the incredulity which they evinced when the statement was made by me. I want to show that it was only the result of want of experience in hydraulic mining, their situation being somewhat like that of the King of Siam, who was offended when an English visitor told him that in his country water became so hard that he could walk on it.

Please let me hear from it at your earliest convenience and believe me to be
Very sincerely yours,
STEPHEN J. FIELD.

In his reply to this petition ex-Senator Fair inclosed the following statements. The first is from Louis Glass:

"At the Spring Valley Hydraulic Gold Mine in Cherokee, Butte county, California, our largest stream was through an 8-inch diameter nozzle under 311 verticle feet verticle pressure, delivered by about a half a mile of two and a half feet diameter iron pipe; and I have seen one of these streams at, say, twenty feet from nozzle, move a boulder weighing about two tons, in a sluggish way, and throw a rock of five hundred pounds as a man would a twenty-pound weight. No man that ever lived could strike a bar through one of

PLACER MINING AND HYDRAULICS. 107

these streams within twenty feet of discharge, and a human being being struck by such a stream would be instantly killed, pounded into a shapeless mass.

"To verify this here is an estimate of power developed under similar circumstances:

"Say 8-inch diameter nozzle 300 feet head, delivered through iron pipe large enough to eliminate friction; 300 feet head by 433 pounds by 50 (square of 8-inch diameter) equals 182,000 pounds aggregate pressure, or 91 tons; but by want of cohesion in the column of water after leaving the nozzle this great force is rapidly dissipated and at about 240 feet the momentum is lost."

The second statement is by Aug. J. Bowie:

"The water which in large hydraulic mines is used under a pressure varying from 200 to 500 feet, is discharged through machines styled 'giants' or 'monitors,' with nozzles from 4 to 9 inches in diameter. Leading up to these nozzles the supply pipe tapers and is lifted to keep the stream from twisting; hence the water as it issues is practically solid.

"A 6-inch nozzle under a 200 feet pressure will discharge 14 cubic feet of water a second, equal to 326 horse-power. The same size nozzle under 450 feet pressure will deliver 21 cubic feet of water per second, which would be equal to a blow of 588,735 foot pounds per second, equivalent to 1070 horse-power. It is absolutely impossible to cut such a stream with an ax or to make any impression on it with any other implement.

"The velocity of the water as it issues from the nozzles would in the cases mentioned vary from 70 to 105 feet per second. The greater the distance from the discharge nozzle the less effective would be the blow; but were a man to be struck by the stream as it comes from the pipe his body would have to resist a continuous force of from 261,000 to 953,000 foot pounds per second, with the result that it would be cut into fragments. There never has been such an accident, but at distances of from 150 to 200 feet men have been killed by very much smaller streams."

It only remains to explain that this tremendous stream tearing away the banks of gravel forces tons of gold bearing dirt through the water-tight open drains known as sluice boxes, which are made of heavy boards covered on the bottom with "riffles" or blocks of stone or wood, with space between them for the gold to settle in.

PLACER MINING AND HYDRAULICS. 109

As the water rushes through, the heavy gold settles in these little spaces over which quicksilver has been sprinkled, and uniting with the quicksilver forms an amalgam. At length the water is turned off with the exception of a gentle stream, the riffle blocks are taken up, the amalgam is scooped out in buckets, and the residue is washed down to the next riffle and so on through the line of sluice boxes. When the water is turned off the workmen take silver spoons to the nail holes or cracks and gather up any gold or amalgam that may have been caught therein. Then come the various processes of breaking up the amalgam, rubbing it and washing it, straining it through canvas or chamois skin, cleaning it by a hot bath in water and sulphuric acid and packing it tightly in the retort, by means of which the quicksilver is all driven off and the pure gold made ready for the assay office.

It may be imagined that the construction of reservoirs to supply water for these great hydraulic monitors is something of an enterprise. As a matter of fact, it involves vast labor and expense. Suitable valleys are selected near the summit of a high range of mountains, huge dams of solid masonary are built across the gorges at the mouths of the valleys, and the melting snows on the surrounding watersheds supply such a reservoir with water, thus storing it until the natural streams have dried up or run so low that they can no longer be of any service. The Sierras with their numerous valleys almost within the line of perpetual snow are especially adapted to this kind of engineering.

The obstacles to be surmounted before processes like this can be made to apply in a country like the Yukon region where the thermometer goes to 65 degrees below zero in the winter and where the ice is broken up for only two months in the year may be imagined. But it is safe to say that where gold is to be found American genius will devise some means of bringing it out within the reach of civilization.

CHAPTER VII.

ALASKA.

It is no unexpected revelation that the soil of Alaska is found to be impregnated with gold. Seward suspected something of the kind when he negotiated the purchase of the territory from the

Russian Government away back in 1867. He was laughed at then for what was termed Seward's folly, and it became quite the fashion for the newspapers of the day to twit the Secretary of State about spending millions of dollars on a stretch of ice and rocks. But Seward never let himself be troubled by the clamor, and he is on record in more than one utterance as declaring that the Alaskan purchase would eventually be found to be the richest portion of the territory of the United States. His phophecy seems about to be fulfilled. Indeed, it has been in the process of fulfillment for many years, and the money which the United States invested in the purchase has already been repaid several times over. The value of the furs alone in the Alaskan territories exceeds by millions of dollars the price paid by Seward for everything. It has been known, too, for many years that the soil was rich in minerals of many kinds. The coal fields are as extensive as any in the world. Copper is known to lie there in vast quantities, and gold has for years been waiting only for the undaunted band of pioneers who were willing to brave the hardships of cold, starvation and travel in their search for the philosopher's stone. Gold has been taken from Alaska before this. The Treadwell Mines, on Douglas Island, have been worked since 1885, and it is now regarded as the most perfectly equipped quartz mining establishment in the world. In 1895 the Director of the Mint reported that gold to the amount of $1,833,733 had been taken from the Alaskan mine and deposited at the United States mints. But quartz mining is not placer mining. It is not the sort of thing that attracts the argonauts, for it requires a great amount of capital and is devoid of the element of romance which renders the gold beds of the Klondike as fascinating to the fortune-seeker as the Californian gold beds were to the fortune-seeker of 1849. A quartz mine is a huge manufacturing establishment with all that is contained in that term, and the profits go to the head of the concern. Placer mining is the field for individual effort, where every man has at least a chance of making a fabulous fortune on his own account. In placer mining one may pick out the gold with his fingers. There is something about that process which appeals to the imagination. And so it happens that while millions of dollars have already been taken out of the Alaskan territory, it remained for the splendid discoveries at Klondike to open the eyes of the world to the surpassing richness of the Alaskan field.

Very few people in the United States, even among the more intelligent and educated classes, fully appreciate the immensity of the territory which was added to the public domain by the purchase of Alaska. The total area of the United States proper, including the fully organized territories, is 2,970,000 square miles. Alaska proper in the mainland contains an area of 580,107 square miles; the islands of Alexander Archipelago, off the southeastern coast, contain 31,205 square miles, and the Aleutian Islands, 6391 square miles. In other words Alaska with its adjacent islands embraces more square miles of territory than twenty-one States of the Union east of the Mississippi River; that is all the New England States, Delaware, Indiana, Kentucky, Maryland, Michigan, Mississippi, New Jersey, New York, North Carolina, Ohio, Pennsylvania, South Carolina, Tennessee, Virginia and West Virginia —State that are represented in Congress by forty- -two Senators and two hundred Representatives. The numerous islands, creeks and inlets of Alaska lengthen out its coast line to 7860 miles, an extent greater than that of the eastern coast line of the United States. Beginning at the southeast the chief creeks and bays are Cook's Inlet, Bristol Bay, Norton Sound and Kotzebue Sound; while, following the same order, the principal headlands, in addition to the extremity of the peninsula, are Cape Newenham and Cape Romanzoff in the Pacific, Cape Prince of Wales in Bering Strait, and Cape Lisbourne, Icy Cape and Point Barrow in the Arctic Ocean. Point Barrow is in 71.23 north latitude, and is the extreme northern point of the country. The territory has an extent of over one thousand miles from north to south, and the Island of Attou, the last of the Aleutian group, is two thousand miles west of Sitka. The longitude of Attou is as many degrees west of San Francisco as Eastport, Maine, is derees east. It is through the possession of Alaska that the American citizen is able to boast that the sun never goes down upon the dominions of the United States. The Governor of Alaska, sitting in his office in Sitka, is very little farther, measuring in a straight line, from Eastport, Maine, than he is from the extreme western limit of his own jurisdiction, which extends beyond the most easterly point of Asia, a distance of nearly one thousand miles, to the one

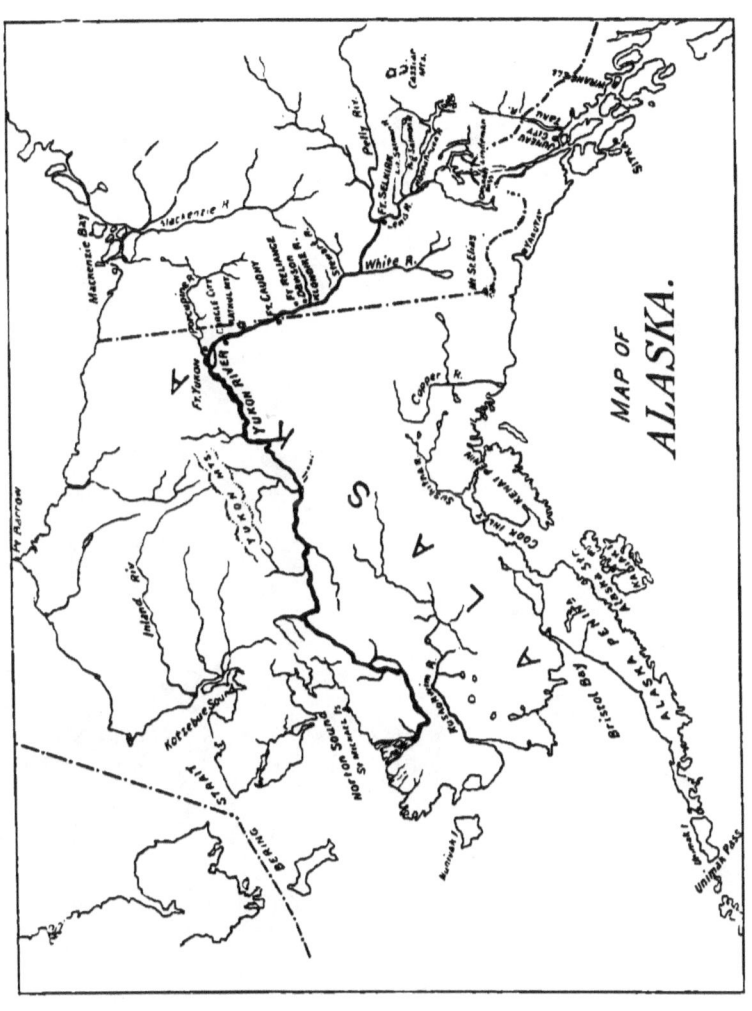

ALASKA.

hundred and ninety-third deree of west longitude, embracing an area very nearly equal to one-fifth of all the States and organized Territories of the Union With its navigable rivers, interminable forests, and lofty mountain ranges, it would be strange, indeed, were it not possessed of natural resources, the development of which is the only condition precedent to the growth of a rich and prosperous State. That these resources are even now comparatively unknown is not to be wondered at in view of the long neglect of the territory by the national government. The extent to which this neglect has been carried is shown by the fact that only since the recent startling reports of the development of the gold region in the interior has the United States seen fit to make any provision for the administration of the law in that part of the territory. It is hardly a fortnight since the office of United States Commissioner for Western Alaska was created by the President, and Charles H. Isham was appointed to the place. Mr. Isham will be stationed at Circle City, but whether he will find any city there upon his arrival is something of a question. He will be authorized to appoint deputy marshals to aid him in enforcing the laws of the

United States. Governor Ryan, the first Assistant Secretary of the Interior, admits that the force employed in the civil government in Alaska is entirely inadequate if there is any appreciable increase at points remote from the towns where government officials are now located.

The gold fields are away up in the Yukon, at the edge of the Arctic circle, hundreds of miles distant from Sitka and other coast towns, where are located the United States Marshals, United States Commissioners, Deputy Marshals and Deputy Commissioners. The active force in the territory that has to carry on a civil government is small. The police force, as it may be termed, consists of a United Statts Marshal and eight Deputy Marshals, eight United States Commissioners and eight Deputy Commissioners. Of course, in case of trouble, the Marshal could exercise the power of a high sheriff and summon the posse comitatus. The United States laws are rigidly enforced in southeastern Alaska along the coast and the citizens of the territory are fully protected in the settlement, but the miners who push several hundred miles beyond civilization will have to be a law unto themselves until other arrangements are made for increasing the civil force of the territory. The general land office has recommended the establishment of two land districts in western Alaska and one of the officers will be located at Circle City. There has been the greatest confusion among the prospectors owing to the absence of facilities for proving up claims, and it is feared that there will be a great number of contentions over mineral land locations in various sections where the gold discoveries have been made. Some of the prospectors who have arrived in San Francisco and Seattle have endeavored to secure government recognition for their claims, only to find that the processes they had gone through with were valueless and that they would be compelled to make the whole wearisome journey over again with witnesses who could testify to their occupation of the land.

The population of Alaska is largely a matter of estimate. According to the latest reports it amounted to about 35,000. Of these about 10,000 might be described as civilized and this number includes not only the whites but the creoles and the Aleutians. Most of these are settled in the southeastern coast country, where the seat of government has been. The people called creoles are descendants three or four generations remote, of a mixed parentage (Russian fathers and native mothers), but it will puzzle even the most learned ethnologist to find anything in their features or complexions by which to distinguish them from the race to which their fathers belonged. They are, to all intents and purposes, white people, fully as intelligent and well informed as would almost any other class of people have been, if subjected to the same wrongs and disadvantages. They, as well as the Aleuts, are civilized people, in the sense that the first were never in a condition of barbarism, while the last, if indeed not fully enlightened, have most certainly been reclaimed from their original savage state. Under the rule of the Russian-American Company the creoles were given the same opportunities for acquiring an education as were afforded to pure blood Russian children, up to a certain age, when they were compelled to enter the employ of the company for a long term of years. The brightest among the creoles and Aleutian boys were carefully trained in navigation, ship building and the mechanical arts, while the girls were taught housekeeping, and thus fitted to become wives of the company's employes, and there are said to be now in the Russian army and navy officers of very considerable rank, and a good many who hold high positions in the civil service of the Empire, who are the progeny of these mixed marriages. The Aleuts are a keen, bright and naturally intelligent people, industrious and provident, the larger portion being educated to a greater or less extent in the Russian language, and that they are well advanced in civilization is evidenced by the fact that they live in comfortable houses, are given to finery in their dress, and are, with scarcely an exception, devout members of one of the Christian churches.

The native Alaskans are a very superior race intellectually, as compared with the people generally known as North American Indians, and are as a rule industrious and provident and wholly self-sustaining. That they yield readily to civilizing influences is evidenced by the fact that wherever the Christian missionaries have gained a foothold, they will be found living in neat comfortable homes of their own construction, and many of them earnestly intent upon bettering their condition, intellectually and morally. They are shrewd and natural-born traders, some are passably good carpenters, and others still are skillful workers in woods and metals. As fast as they can obtain employment from the white men at reasonable wages (and the most ignorant among them know the value of their labor) they abandon the chase and the fishing grounds, and serve their employers faithfully so long as they are well treated. At least a hundred are employed at the great mine and mill on Douglas Island, and as laborers and miners are far superior to the Chinese.

Of course, with the influx of miners to the new placer diggings the population of whites will be greatly increased, and it is certainly not a rash estimate that the total population of the territory will be more than doubled in the next twelve months. So far as is known there are about three thousand white men now scattered over the gold fields, and most of these have been concentrated about the Klondike region. Five thousand more are on the way, and with the opening up of spring they will begin to pour in upon the unexplored country by the thousands. With the rapid increase of population and the direction of attention to the new Eldorado there it will only be a short time before transportation facilities are afforded between Juneau and the gold fields and the way paved for establishing the beginnings of a great Commonwealth.

It appears now that Juneau, situated as it is at the head of tidewater and at the gateway to the gold country, will be the most important city of Alaska. Indeed, it is already the metropolis of the Territory, although Sitka still remains the capital, and, owing to its age and its situation, will continue to be an important point. The population of Sitka, in the latest reports, was about 1200. Juneau is destined to be the outfitting point for all miners on their way to the Yukon gold fields. It has a population of nearly fifteen hundred, which is bound to rapidly increase. It is more nearly than other Alaskan city on a par with the cities farther south. It is the headquarters of several steamboat lines, has a city hall and court house, substantial walls, water works, electric lights, hotels and a large number of fine buildings. It is a picturesque city, situated at the foot of the mountains, which are snow-capped throughout the year and down which avalanches are constantly tearing. One or more avalanche rushes down the mountain side every day, and these incidents lend to life there an interest peculiarly its own.

It is a singular circumstance that glaciers approach nearer to the ocean here at Juneau than at any other place in the world. Indeed it is the only place so far as known where glaciers come near to the ocean at all, but here the approach is so close and the motion oceanward is so steady that the waters around the city are filled with floating icebergs, somewhat to the peril of seafaring men.

Juneau was founded in the winter of 1880 and six months after the discovery of gold (August 15, 1880) by Joseph Juneau and Richard Harris. It went under the name of Harrisburg at first and afterwards was called Rockwell, but the miners at a meeting about a year after its foundation decided to rechristen it in honor of the discoverer of gold. Within a year it has become a flourishing mining town, and now it is the commercial centre of Alaska. It supports three weekly newspapers.

The exploration of the northern coast was chiefly the work of the British navigators, Cook, Beechy and Franklin, and of the officers of the Hudson Bay Company. The principal river of Alaska is the Yukon, which rises in British America, and, receiving the Porcupine River at Fort Yukon, flows westward across the territory and falls into the Pacific Ocean to the south of Norton Sound. At a distance of 600 miles from the sea this magnificent river has a width of more than a mile. Its tributaries would in Europe be reckoned large rivers, and its volume is so great than ten miles out from its principal mouth the water is fresh. Among the other rivers of Alaska are the Copper River, the Suschitna, the Nuschagak and the Kuskokwim, falling into the Pacific, and the Colville, flowing northward into the Arctic Ocean. A great mountain range extends from British Columbia, in a northwest direction, along the coast of Alaska, the summit being covered with snow and glaciers. Mount St. Elias, an active volcano, in 60.18 north latitude and 140.30 west longitude, rises to the height of 14,970 feet above the sea. The mountain chain runs along the peninsula, which has given its name to the country, and at the western extremity there are several volcanic cones of great elevation, while in the Island of Uminak, separated from the mainland by only a narrow strait, there are enormous volcanoes, one rising to more than 8000 feet in height. In the interior and to the north the country is also mountainous, with great intervening plains.

The northwest coast of this part of America was discovered and explored by a Russian expedition under Behring in 1741, and at subsequent periods settlements were made by the Russians at various places, chiefly by the prosecution of the fur trade. In 1799 the territory was granted to a Russo-American fur company by the Emperor Paul VIII, and in 1839 the charter of the company was renewed. New Archangel, in the Island of Sitka, was the principal settlement, but the company had about forty stations. They exported annually 25,000 skins of the seal, sea-otter, beaver, etc., besides about 20,000 sea-horse teeth. The privilege of the company expired in 1863, and in 1867 the whole Russian possessions in America were ceded to the United States for a money payment of $7,200,000. The treaty was signed March 30 and ratified on June 20, 1867, and on October 9 following the possession of the country was formerly made over to a military force of the United States at New Archangel (now Sitka). Portions of Alaska were explored in 1859 by the employes of the Russo-American Telegraph Company in surveying a route for a line of telegraph which was destined to cross from America to Asia near Behring Strait—a project which was abandoned, after an expenditure of $3,000,000, on communication with Europe being secured by the Atlantic cable.

The government of Alaska lies in a Governor, who is appointed by the President. It has not yet a full territorial form of government.

The climate of the Alaskan coast regions is much milder, even in the higher latitudes, than it is in the interior, or in corresponding latitudes on the Atlantic coast. This is easily explained and understood when the natural forces productive of this milder temperature are contemplated.

The most important among them is a thermal current resembling tht Gulf Stream in the Atlantic. This current, known as the Japanese or Kuro Siwo, has its origin under the equator near the Molucca and Philippine Islands, passing northward along the coast of Japan, and crosses the Pacific to the southward of the Aleutian Islands, after throwing a branch through Bering Sea, in the direction of Bering Strait. The main current strikes the coast of British Columbia, where it divides again, one branch turning northward toward Sitka, and thence westward to the Kadiak and Shumagim Islands.

The comparatively warm waters of these currents affect the temperatures of the superjacent atmosphere, which, absorbing the latent heat, carries it to the coast with all its mollifying effect. Thus the oceanic and atmospheric currents combine in mitigating the coast climate of Alaska, and this process is greatly aided by the configuration of the extreme northwestern shores of the Pacific, backed as they are with an almost impenetrable barrier of lofty mountains, which holds back from the interior the warm, moist atmospheric currents coming in from the ocean, deflecting at the same time the ice-laden northern gale from the coast to the interior.

To Hon. A. P. Swineford, who was Governor of Alaska in 1886, belongs the distinction of having first emphatically called the attention of the United States Government to the splendid possibilities of Alaskan development. In the very first report which he made to the Secretary of the Interior in October, 1885, he declared that the natural resources of Alaska, as yet in the infancy of their development, were such as might be made, in the near future, a most important addition to the aggregate wealth of the nation.

"I have seen enough to convince me," he said, "that no other Territory of the Union, at so early a period in its civil history, presented nearly so many or as great possibilities for the future. That Alaska was not supplied with local civil government a dozen years ago is to be deplored; that so-called scientists in the pay of the General Government have heretofore 'damned with faint praise,' if they did not openly condemn the country as utterly worthless, save for its valuable fur trade—basing their statements on what they were able to see, looking at its rugged coast from their favorite standpoint of the Prybilov Islands—is still more to be regretted, for the reason that the tardy and at last only partially performed act of justice on the one hand was but the result of either the ignorant or willful misstatements of those to whom Congress looked for information upon which to base any and all legislation affecting the rights, privileges and interests of Alaska and its people.

"Nowhere in my home travels, from Lake Superior to the Gulf of Mexico, from Washington to Sitka, have I seen a more luxuriant vegetation than in Southeastern Alaska. I find the hardier vegetables all growing to maturity and enormous size; white turnips weighing ten pounds, cabbages twenty-seven pounds, and as fine potatoes as can be found in any of the Eastern markets I found growing at Wrangell, Juneau and in Sitka. Wild timothy and red-top grow to a height of from five to seven feet, and in the vicinity of Sitka all the hay was cured during the past summer that will be required during the winter, and I am satisfied, from personal observation, that hundreds of tons more could have been harvested. The few cattle I have seen are sleek and in the best possible condition, and I unhesitatingly give it as my opinion that the country is well enough adapted to grazing purposes to render wholly unnecessary the importation of beef, even when the population of the Territory shall have grown far beyond the number requisite to its admission as a State."

As an indication of the difficulty Alaska has had in receiving recognition according to its true worth executive document No. 36 of the House of Representatives, second session, Forty-first Congress, may well be quoted here. It contains the report of a special agent of the Treasury Department on the subject of Alaska. From it the following passages are taken:

"The price paid for the Territory, $7,200,000, is but a small item of its cost to the United States. Provided the public debt be paid within twenty-five years, annual interest on the purchase money at the rate of six per cent. would in that period amount to $23,701,792.14, which, added to the principal, would make the total cost of the Territory $30,901,792.14. To this sum there must be added the expense of the military and naval establishments, say $500,000 per annum, or $12,500,000 in twenty-five years, which is a much smaller estimate than can be predicted on the expenditure of the last two years, resulting in a grand total cost on the above basis of $43,401,-792.14.

"In return for this expenditure we may hope to derive from the seal fisheries, if properly conducted, from $75,000 to $100,000, and from customs $5000 to $10,000 per annum, a sum insufficient to support the revenue department, including the present expensive cutter service attached to the district; nor can we look for any material increase of revenue for many years, except in the event of extraordinary circumstances, such as the discovery of so large deposits of minerals as would produce an influx of population.

"As a financial measure it might not be the worst policy to abandon the Territory for the present, until some possible change for the better shall have taken place, but for political reasons this course may not be advisable."

Notwithstanding the above calculations and predictions the management of the Seal Islands alone paid into the United States Treasury between $6,000,000 and $7,000,000 in rental and royalties within twenty years, independent of the "extraordinary circumstances" referred to by this special agent. It is safe to assert that, since the system of leasing the Prybilov Island was inaugurated within a few weeks of the date of the report quoted here and up to the expiration of the first term of twenty years, the revenue covered into our Treasury from Alaska has always exceeded the expenditures, while as a factor in the internal commerce of the United States, and especially of our Pacific coast, Alaska has assumed a position of considerable importance.

A better understanding of the advantages de-

GOLD-BEARING QUARTZ AT YAKUTAT.

rived by the country at large by the purchase of Alaska can be obtained by perusing the subjoined statement of products of the Territory since it came into our possession. The statement embraces only the principal articles of export, and can be relied upon as being conservative and within actual limits of Alaska's products:

VALUE OF PRODUCTS OF ALASKA FROM 1868 TO 1890.

Furs	$48,518,929
Canned salmon	9,008,497
Salted salmon	603,548
Codfish	1,246,650
Ivory	147,047
Gold and silver	4,631,840
Total	$64,156,511

Products of the whaling industry:

Whale oil	$2,853,351
Whalebone	8,204,067
Total	11,057,418

Aggregate	$75,213,929.

CHAPTER VIII.

QUARTZ MINING IN SOUTHEASTERN ALASKA.

This handbook would not approach completion if it refrained from a description of the wonderfully productive gold mines which have been worked in southeastern Alaska for the past twelve years, and which in 1895 contributed nearly $2,000,000 to the gold supply of the world. These quartz mines are the most perfectly developed in the world, and are increasing in productiveness every year. The gold yield of Alaska in 1894 was $1,288,334. In 1895 it increased to $2,328,419.

For 1895 the yield of the quartz mines on Douglas and Unga Islands alone equaled the entire product of the territory the years before, without counting the other mining fields which have been more fully developed.

During the year 1895, 300 stamps were dropping on Douglas Island and during the summer 125 stamps were dropping on the mainland.

Other outlying districts are also coming into prominence, mainly on Admiralty Island, upon which a new ten-stamp mill is now ready for running, being operated by the Alaska-Willoughby Gold Mining Company. On Unga Island some very extensive and productive quartz operations are being carried on.

In southeastern Alaska, so far, all the placer mining has been done in gravel deposits, which were made auriferous by the wash from quartz veins.

The distinction of the first discovery of gold in that extensive and important mining region of which the town of Juneau is the centre, is shared by two pioneer prospectors, Richard Harris and Joseph Juneau. In the summer of 1880 these men started in a canoe from the quaint old town of Sitka to prospect the mainland coast, and about August 15 discovered gold in a stream which they aptly named Gold Creek. Their stock of provisions being nearly exhausted, they did not ascend the stream to its source and soon returned to Sitka, taking with them 150 pounds of gold quartz and 13 grains of "dust." Having secured another outfit, they hurried back to Gold Creek, and soon found its source in a little round valley inclosed by steep, glacier-capped mountains. This spot they named Silver Bow basin, after a place of that name in Montana. On the mountain slopes, encircling the basin, gravel was found worth from 15 to 30 cents a pan, and quartz that seemed to have been splashed with gold. October 4 Juneau and Harris, with the aid of three natives, located their choice of the placer ground, and within a month located 18 quartz claims, organized Harris mining district, adopted local rules for the new district, and staked off a town site near the mouth of Gold Creek, which they named Harrisburg. They then returned to Sitka with 960 pounds of gold ore, worth $14,000.

This golden cargo crazed the quiet town, and a number of adventurous fellows, procuring boats, canoes, or steam launches, rushed off to the new diggings with Juneau and Harris. The season was too far advanced for prospecting in the basin, so log cabins were built on the site staked off by the founders of the camp. During the winter of 1880-1881 the town of Harrisburg flourished; five general merchandise stores were established and saloons appeared so quickly as to seem spontaneous; miners and frontiersmen generally flocked in from Wrangell and British Columbia, add all waited impatiently for spring. At a miners' meeting in February, 1881, the town name was changed to Rockwell, in honor of Lieutenant

THE YUKON FLATS.

Rockwell, United States Navy, and the following November, at another meeting, the place was rechristened Juneau, in honor of Joseph Juneau. On the 27th of January John Pryor, Antone Marx, Frank Berry, James Rosewald and William Mehan discovered placer and quartz on the beach of Douglas Island, four miles from the town. They began working the placers early in March, washing out 27 ounces of gold in the first three days' work.

The first shipment of gold from the new camp was taken from this claim and amounted to 84 ounces. The claim, still known as Ready Bullion, yielded about $12,000 in 1881, $3000 in 1882 and in 1884 was sold to John Treadwell. These are the beginnings of the famous Treadwell mines, from which enough ore has been taken out in the last ten years to pay the purchase of Alaska and more.

An expedition of great value in the exploration of the gold resources of southern Alaska was undertaken under the auspices of the United States Geological Survey in 1895, under the direction of Dr. George F. Becker. Dr. Becker was assisted by Mr. C. W. Purington, who devoted himself especially to the examination of the gold deposits, and associated with him was Dr. W. H. Dall, the Alaskan authority, who had immediate charge of an examination of the coal resources.

The instructions to the party were to examine the gold and coal deposits in the vicinity of the shore line and islands along the coast of the territory, and not to attempt to penetrate into the interior.

Dr. Becker and Mr. Purington examined the Treadwell mine, on Douglas Island, and found that the mine was in slates of sedimentary origin, probably of Triassic age, and that it had been penetrated by a heavy dike of diorite or tonalite and by two other intrusive masses. The last of these is a rock of basaltic character, and its eruption seems to have occurred at the same time as the mineralization. Both the diorite and the slate were ruptured along a zone which is at some points several hundred feet in width, and the interstitial spaces have been filled with ore. In the diorite the masses were in great part reduced to fragments, and these have been decomposed and impregnated. In the slate the fractures mostly followed the cleavage, and the deposit there assumes the form of a "stringer lead." The claims to the southward of the Treadwell are controlled by the same company, and are profitable, but the next claim to the northward is said to be too poor to pay. The ore of Treadwell averages only $2.50 to $3 per ton, but, owing to the enormous scale of the workings, there is a large prfit in working it.

The Silver Bow basin lies about three miles north of east of Juneau. A considerable number of small veins of rather rich ore occur in the southern side of the basin. The basin was formerly occupied by a large glacier. After the retreat of the glacier the basin was occupied by a lake, and the lake beds are successfully worked for gold by the hydraulic process.

Sheep Creek basin is separated from Silver Bow basin by a divide, and the same series of quartz veins extend into it. About 55 miles to the southeast of Juneau, at Sumdum, there is a very promising vein which is yielding good bullion, although the property is only just being developed. At Seward City, near Berners Bay, about 50 miles north of Juneau, there are also veins which are extremely rich at some points and are yielding gold. On Admiralty Island, about 30 miles from Juneau, there are promising veins, on which it is expected that mining will be commenced during the summer of 1896. Near Sitka, especially along Silver Bay and in the country to the southeast of it, there are numerous veins, some of which have yielded a little gold. The conditions do not warrant an opinion as to their future.

At Yakutat Bay, just to the eastward of Mount St. Elias, there has been some beach mining, as there has also been along the west shore of Kadiak Island. The ease of working and the unlimited supply of sand make beach mining on the western coast of North America very attractive, but the capriciousness of the distribution of pay streaks and the difficulty of saving the gold commonly rob such undertakings of success. The amount of gold which occurs in this manner in the sand is enormous, but as yet there are few if any reliable records of large profits having been made from beach mines, either in Alaska or to the southward.

On Kadiak Island, in Uyak Bay, there are several promising-looking gold-quartz veins, 2 feet or so in thickness, upon which prospecting is now going on. Stream gravels are also being worked on Turn-again Arm, at the head of Cook Inlet. The only successful working was on Bear Creek, but the capriciousness of the distribution of pay Becker could obtain the average results were not more than $5 per day per man. A later report, received after Dr. Becker's visit, is that richer gravel has been discovered near the head of Turnagain Arm.

The island of Unga is in the Shumagin Archipelago, about a thousand miles south of west from Sitka. Near Delaroff Bay, on this island, is the Apollo Consolidated mine, which is now yielding at the rates of over $300,000 a year. The ore occurs in interstitial spaces in a crushed zone of andesite. It averages between $8 and $9 per ton, much of the gold being free, though heavy bunches of sulphurets are of frequent occurrence in it.

Although auriferous quartz has been found on the island of Unalaska, nothing like a mine has yet been discovered.

Speaking of the gold mines of Southeastern Alaska as early as 1886 Governor Swineford said:

"The extensive reduction works on Douglas Island, opposite Juneau, are, perhaps, the most complete of any to be found on the Pacific slope.

They are supplied with twenty-four batteries of five stamps each, with all the necessary machinery and appliances for the extraction of the free gold, and chlorination works for the treatment of the sulphurets. During July and August the mill, running to not much more than half its full capacity, turned out $115,000 in gold bullion, while the accumulated sulphurets (concentrates) awaiting treatment were shown by frequent assays to be worth not less than $100,000 more.

"Since the middle of September the mill has been running to its full capacity, and a personal examination of the mine from which it is supplied with ore leads me to confidently expect very much better results from this time forward. The mine itself is located in what appears to be simply a great mountain of gold-bearing quartz. Into this immense repository of the precious metal a tunnel has ben driven to a length of nearly, if not quite, 500 feet, as nearly as I could judge, at right angles with the trend of the ledge, and on a level at least 250 feet below the outcrop on which the miners were at work breaking and milling the rock down through a winze to the tram-cars in the tunnel. A careful examination of the tunnel reveals well defined foot and hanging walls, very nearly 400 feet apart, between which nothing but the same kind of rock as that being milled at the time of my visit can be seen on either side. The rock is what is called "low grade milling," carrying free gold and sulphurets, and yields an average, I am told, of about $8 per ton. No selection of the rock is necessary, everything from between the walls going to the stamps. It is truly a phenomenal deposit and the mine one that promises to figure more largely in the mining history of the world than any other of which we have any record.

"In the rear of Juneau two or three miles, on the mainland, is Silver Bow Basin, where some rich placer mines are being worked, but thus far I have not been afforded an opportunity to visit or examine them. The value of the product of these mines, however, has been estimated by well-posted persons at not less than $150,000 in 1884, and the opinion prevails that the shipment of "dust" will be much larger the present year. I noticed while in Juneau that most of the traders were buying gold dust, and was told that many of the miners in the basin were doing well, and some of them amassing comfortable fortunes.

"In the absence of other discoveries it would yet be hardly probable that the gold-bearing ledges and basins of the Territory should be confined to this one particular locality. Fortunately there is abundant evidence going to show that the developments at Juneau are but the precursors of others yet in abeyance, and which await only the application of similar effort in the way of the expenditure of labor and capital to make them profitably productive. In the near vicinity of Sitka there are promising ledges, one of which has been wrought for years in a desultory way by a single prospector, who, doing only the assessment work required by the mining law, has yet been able to support himself and family from the proceeds extracted from his incipient mine by the most primitive appliances—principally an ordinary hand pestle, mortar and pan. While there can be little doubt of the existence of gold along the coast range of mountains, and on many of the islands of the Alexander Archipelago, the geological formation and general characteristics of which appear to be identical with those of the mainland, the work of development will necessarily proceed slowly as compared to the prog-

ST. MICHAEL ISLAND.

ress made in the other mining districts of the United States, owing to the difficulties which beset the path of the prospector, unless, indeed, convenient access to tidewater may wholly or in part be found to counterbalance the disadvantages of high and precipitous mountains, covered with a dense growth of timber, underbrush and fallen trees, with two or three feet of intertwining, closely woven vines and moss covering the ground itself, and which will obstruct and render more than usually difficult the work of exploration, though not necessarily an obstruction in the way of subsequent mining operations. The difficulties mentioned will, however, be partially obviated by the first discovery in any particular locality, which will serve as a starting point from which to prosecute explorations with a better knowledge of the formation, and, consequently, with much less labor and expense. In addition to the compensating advantage of contiguity to navigable waters there is unlimited water power for the operation of mining and milling machinery and an abundance of timber for all purposes."

CHAPTER IX.

THE WONDERFUL YUKON COUNTRY.

Although the eyes of the world are only just beginning to be opened to the surpassing interest of the immense area of country watered by the Yukon River, there are men living to whom the marvelous features of that great stretch of country are no new thing. The highest authority on all questions pertaining to the Yukon is Dr. W. H. Dall, of the Smithonian Institution, in Washington, who more than a generation ago went up into that country with the Western Union expedition sent out to survey for the proposed Russian-American telegraph line and who has made several journeys to the same region since. Dr. Dall embodied the observations of his early visits in a book published in 1870, entitled, "Alaska and its Resources," which is easily the most comprehensive work issued on the general subject of our Alaskan possessions. No subsequent explorers have succeeded in fully replacing it, although the latest census reports are very complete, considering the difficulties of exploration.

THE WONDERFUL YUKON COUNTRY.

What the Amazon is to South America, the Mississippi to the central portion of the United States, the Yukon is to Alaska. It is a great inland highway, which will make it possible for the explorer to penetrate the mysterious fastness of that still unknown region. The Yukon has its source in the Rocky Mountains of British Columbia, and the Coast Range Mountains of southeastern Alaska, about 125 miles from the city of Juneau, which is the present metropolis of Alaska. But it is only known as the Yukon River at the point where the Pelly River, the branch that heads in British Columbia, meets with the Lewis River, which heads in southeastern Alaska. This point of confluence is at Fort Selkirk, in the Northwest Territory, about 125 miles southeast of the Klondike. The Yukon proper is 2044 miles in length From Fort Selkirk it flows northwest 400 miles just touching the Arctic circle; thence southward for a distance of 1600 miles, where it empties into Bering Sea. It drains more than 600,000 miles square of territory, and discharges one-third more water into Bering Sea than does the Mississippi into the Gulf of Mexico. At its mouth it is sixty miles wide. About 1500 miles inland it widens out from one to ten miles. A thousand islands send the channel in as many different directions. Only natives who are thoronghly familiar with the river are trusted with the piloting of boats up the stream during the season of low water.

Even at the season of high water it is still so shallow as not to be navigable anywhere by seagoing vessels, but only by flat-bottomed boats with a carrying capacity of four to five hundred tons.

The Yukon River is absolutely closed to travel save during the summer months. In the winter all approaches are locked with impenetrable ice and the summer lasts only from ten to twelve weeks, from about the middle of June to the early part of September. Then an unending panorama of extraordinary picturesqueness is unfolded to the voyager. The banks are fringed with flowers carpeted with the all pervading moss or tundra. Birds, countless in numbers and of infinite variety of plumage, sing out a welcome from every tree top. Pitch your tent where you will be in midsummer, a bed of roses, a clump of poppies and a bunch of blue bells will adorn your camping. But high above this paradise of almost tropical exuberance, giant glaciers sleep in the summit of the mountain wall, which rises up from a bed of roses, has disappeared before icy breath of the Winter King, which sends the thermometer down to eighty degrees below the freezing point.

The Lewis River is the best known of the tributaries to the Yukon, having been used for the past twelve years as the highway from Southeastern Alaska to the gold diggings on the Yukon. Its length from Lake Lindeman, one of its chief sources, to the junction with the Pelly is about 375 miles, and it lies entirely in British territory, with the exception of a few miles of the lake at its head.

The Pelly River takes its rise about Dease Lake, near the headwaters of the Stkine River, with a length of some 500 miles before joining the Lewis to form the Yukon River. The union of these two streams forms a river varying from three-quarters of a mile to a mile in width. For many miles on the northern bank is a solid wall of lava, compelling a swift current to follow a westerly course in search of an outlet to the north. The southern bank is comparatively low, formed of sandy, alluvial soil. A few miles above the White River the stream takes a northerly course through a rugged, mountainous country, receiving the addition of the waters of the White River on the south, so called from the milky color of its water, and a few miles farther on the waters of the Stewart on the north. The current is exceedingly swift here, especially at a high stage of water, being at least six or seven miles an hour. From Stewart River to Fort Reliance both banks are closed in by high mountains, formed chiefly of basalt rock and slaty shale. Many of the bluffs are cut and worn into most picturesque shapes by glacial action. At Fort Reliance, an abandoned trading post, the general course of the stream changes to northwest, continuing thus for a distance of about 500 miles, or as far as the confluence with the Porcupine River, which flows from the north.

Some forty miles from Fort Reliance the mouth of Forty Mile Creek is passed, where is located the miners' trading post and where for some time were found the chief gold diggings. Some thirty-eight miles from there the river crosses the eastern boundary of Alaska. For a hundred miles after crossing the boundary the river runs in one broad stream, confined on either side by high banks and a mountainous country, known as the "Upper Rampart." It

THE WONDERFUL YUKON COUNTRY 149

then widens out, and for a distance of 150 miles is a network of channels and small islands. At Old Fort Yukon, an abandoned Hudson Bay post, it attains its high northern latitude, being just within the Arctic circle. From main bank to bank the distance has been found to be exactly seven miles at a point just above the site of Fort Yukon. This place is probably the only serious obstacle to navigation that is met with from the mouth of the river to Fort Selkirk, a distance of over one thousand six hundred miles, the channel here shifting from year to year, and at certain stages of water it is difficult to find. From Fort Yukon to the mouth the river has been frequently traveled and well described.

According to Dr. Dall the character of the country in the vicinity of the Yukon River varies from low, rolling and somewhat rocky hills, usually easy of ascent, to broad and rather marshy plains, extending for miles on either side of the river, especially near the mouth. There are no roads, except an occasional trail, hardly noticeable except by a voyageur. The Yukon and its tributaries form the great highways of the country.

The soil is usually frozen at a depth of three

or four feet in ordinary situations. In colder ones it remains icy to within eighteen inches of the surface. This layer of frozen soil is six or eight feet thick. Below that depth the soil is usually destitute of ice.

This phenomenon appears to be directly traceable to want of drainage, combined with non-conductive covering of moss, which prevents the scorching sun of the boreal midsummer from thawing and warming the soil.

A singular phenomenon on the shores of Escholtz Bay, Kotzebue Sound, was first observed and described in the voyage of the Rurik by Kotzebue and Chamisso, and afterwards in the appendix to the voyage of the Herald by Buckland and Forbes.

It consists of bluffs or banks (thirty to sixty feet high) of apparently solid ice, fronting the water, which washes on a small beach formed by detritus, at the foot of the bank. These continuous banks of ice, strange to say, are covered with a layer of soil and vegetable matter, where, to use the words of the renowned botanist, Dr. Seemann, "herbs and shrubs are flourishing with a luxuriance only equaled in more favored climes."

The climate of the Yukon Territory in the interior (as is the case throughout Alaska) differs from that of the sea coast, even in localities comparatively adjacent. That of the coast is tempered by the vast body of water contained in Bering Sea, and many southern currents bringing warmer water from the Pacific, making the winter climate of the coast much milder than that of the country, even thirty miles into the interior; this, too, without any high range of mountains acting as a bar to the progress of warm winds. The summers, on the other hand, from the quantity of rain and cloudy weather, are cooler and less pleasant than those of the interior. The months of May and June, however, and part of July, are delightful—sunny, warm and clear. To quote Seemann again, on the northern coast, "the growth of plants is rapid in the extreme. The snow has hardly disappeared before a mass of herbage has sprung up, and the spots which a few days before presented nothing but a white sheet are teeming with active vegetation, producing leaves, flowers and fruits in rapid succession." Even during the long Arctic day the plants have their period of sleep, short, though plainly marked, as in the tropics, and indicated by the same drooping of the leaves and other signs, which we observe in milder climates. The following table shows the mean temperature of the seasons at St. Michael's, on the coast of Norton Sound, in latitude 63 degrees 28 minutes; at the Mission, on the Yukon River, one hundred and fifty miles from its mouth, in latitude 61 degrees 47 minutes; at Nulato, four hundred and fifty miles farther up the river, in latitude 64 degrees 40 minutes (approximate), and at Fort Yukon, twelve hundred miles from the mouth of the river and about latitude 66 degrees 34 minutes:

Means for	St.Mich's.	Mission.	Nulato.	Ft.Y'n.
Spring	29.3	19.62	29.3	14.22
Summer	53.0	59.32	60.0	59.67
Autumn	26.3	36.05	36.0	17.37
Winter	8.6	0.95	14.0	23.80
Year	29.3	26.48	27.8	16.92

The mean annual temperature of the Yukon Territory, as a whole, may be roughly estimated at about 25 degrees. Open water may be found on all the rivers in the coldest weather and many springs are not frozen up throughout the year.

At Fort Yukon Dr. Dall says he has seen the thermometer at noon, not in the direct rays of the sun, standing at 112 degrees, and he was informed by the commander of the post that several spirit thermometers graduated up to 120 degrees had burst under the scorching sun of the Arctic midsummer, which can only be thoroughly appreciated by one who has endured it. In midsummer on the Upper Yukon the only relief from the intense heat, under which the vegetation attains an almost tropical luxuriance, is the brief space during which the sun hovers over the northern horizon.

The rain fall is much greater in summer on the coast than in the interior. The months of May, June and part of July bring sunny, delightful weather; but the remainder of the season, four days in a week at least, will be rainy at St. Michael's. October brings a change. The winds, usually from the southwest from July to the latter part of September, in October are mostly from the north, and, though cold, bring fine weather. They are interrupted occasionally by gales, the most violent of the season, from the southwest; piling the driftwood upon the shores, where it lies until the succeeding fall, unless carried off by the natives for fuel.

The valley of the Lower Yukon is somewhat

foggy in the latter part of the summer; but as the river is ascended the climate improves and the short summer at Fort Yukon is dry, hot and pleasant, only varied by an occasional shower.

The first requisite for habitation or even exploration in any country is timber. With it almost all parts of the Yukon territory are well supplied. The treeless coasts even of the Arctic Ocean can hardly be said to be an exception, as they are bountifully supplied with driftwood from the immense supplies brought down by the Yukon, Kuskoquim and other rivers, and distributed by the waves and ocean currents.

The largest and most valuable tree found in this district is the white spruce, which is found over the whole country a short distance inland, but largest and most vigorous in the vicinity of running water. It attains not unfrequently the height of fifty to one hundred feet, with a diameter of over three feet near the butt; but the most common size is thirty or forty feet and twelve to eighteen inches at the butt. It is quite durable. Many houses, twenty years old, built of this timber, contained a majority of sound logs, but when used green, without proper seasoning, it will not last over fifteen years. These trees decrease in size, and grow more sparingly near Fort Yukon, but are still large enough for most purposes.

The tree of next importance in the economy of the inhabitants is the birch. This tree rarely grows over eighteen inches in diameter and forty feet high.

Several species of poplar abound, the former along the water side and the latter on drier uplands. The first mentioned species grows to a very large size, frequently two or three feet in diameter and forty to sixty feet high. The timber, however, is of little value, but the extreme softness of the wood, is often taken advantage of by the natives with their rude iron or stone axes, to make small boards or other articles for use in their lodges. They also rub up with charcoal the down from the seed-vessels for tinder.

Willows and alders are the most abundant of trees and all sizes of the former may be found.

The treeless coasts of the territory, as well as the lowlands of the Yukon, are covered in spring with a most luxuriant growth of grass and flowers. Among the more valuable of these grasses is the Kentucky blue grass, which grows as far north as Kotzbue Sound, and on the coast of Norton Sound with a truly surprising luxuriance. It reaches in very favorable situations four or even five feet in height and averages at least three feet.

Grain has never been sown to any extent in the Yukon territory. Barley was once or twice tried at Fort Yukon in small patches and succeeded in maturing the grain, though the straw was very short. Attempts were also made at 40-Mile Post in 1890.

Turnips and radishes always flourished extremely well at St. Michael's and the same is said of Nulato and Fort Yukon. Potatoes succeeded at the latter place, though the tubers were small They were regularly planted for several years, until the seed was lost by freezing during the winter. At St. Michael's they did not do well.

Salad was successful, but cabbage would not head. The white round turnips, grown at St. Michael's, from European seed, were very large, some weighing five or six pounds.

The inhabitants of the upper Yukon, from the Rampart house to the boundary, formed part of the nation known to the English missionaries of the Hudson Bay side as the Tukudh Indians, tribes of which extend over the country enclosed by the Porcupine River, the Peel River to the MacKenzie, the Upper Yukon to the neighborhood of the Stick Indians in the south and to the southeast in the McMillan River country. They speak of themselves, however, as Yukon Indians. Their language has been put into print by the venerable Robert McDonald, archdeacon, Bibles and hymn books being universally read by all from Nuklukayet up. They are of average size, lithe and active, many of them being quite graceful in their carriage. In appearance they approach the typical North American Indians; sharp features, aquiline nose, and high cheek bones, with very small feet and hands. They are nomadic in their ways of life, living in temporary camps both winter and summer, either in the mountains or on the river banks, according to the habits of the game they are hunting.

Some few in the neighborhood of the mining camp are perceptibly changing their mode of life. Around the trading post at Forty Mile Creek there are a number of log cabins built and inhabited by them the year around, and they fully apprecite the advantages of stoves and clothing from the States. The younger men are said to be more fastidious in their dress than the average white man. They are industrious and fairly enterprising, many of them working successfully at mining for wages paid by the whites, and some are mining on their own account. They make excellent boatmen, poling a boat with skill, boats built of sawed lumber being preferred for river navigation to their own birch canoes. Docile and peaceable both among themselves and with the miners, they are strongly imbued with the teachings of the English missionaries, with whom they had more or less intercourse for many years previous to occupation of the country by the United States. Formerly their chief subsistence was cariboo and moose meat, and fish they only knew during the summer and fall, but since the arrival of the miners they depend each year more and more on white men's provisions. Obtaining pay for work, they also avoid the necessity of hunting for fur to buy provisions with, as used to be the case in former years; hence the falling off of the supply of furs from that section.

The population is very sparse. At certain times during the year a traveler might pass down the Yukon from Forty Mile Creek to Nuklukayet and hardly see a score of natives in a distance of 800 miles. The different villages or communities seem to be under the guidance of chiefs and sub-chiefs, though there does not appear to be much authority exerted by them.

Their mode of transportation in summer time is by rafts, boats and birch canoes, and is entirely confined to the streams and water courses; in the winter time sleds are used, drawn by dogs, men or women. Their language is known to the missionaries as a dialect of Tukudh (Tukuth), but they converse with the traders in a jargon called "Slavey," a mixture of Canadian French and hybrid words of English, something in the nature of the "Chinook" of southeastern Alaska.

At Nuklukayet and down to the vicinity of Nulato changes are to be observed in the natives; though very similar in general appearance, they seem to be a mixture of tribes from the Koyukuk and Tanana Rivers and of Ingalik, from lower down the Yukon.

Their language is different, though many can converse in a dialect that is understood by the Upper Yukon people. They are not so nomadic in their way of life, living in villages, building log cabins and huts of earth and logs. They depend most largely on the supply of fish and not so much on game. They are mostly addicted to paganism, being more superstitious and depending on instructions from the shaman, or medicine man. They also are becoming yearly more dependent on provisions from the States, but have to procure them by trapping fur-bearing animals to a far larger extent than those of the upper river. They are shrewd traders, taking advantage of every point. They do not so readily adapt themselves to the ways of the white man. They are more pugnacious, quick-tempered, resenting a fancied injury or insult very quickly with force. Many years ago some of them killed a white woman, the wife of a trader at a post a few miles up the Tanana River, at the instigation of a shaman. Four years ago at Nuklukayet, on account of a disagreement with a trader, they broke open the store, scattered the goods about recklessly and would have shed blood if they had not met with adequate resistance. Religious teaching does not seem to have the same effect upon them as on the natives on the upper river. They have had visits from Russian priests and resident English missionaries in past years, without much notable effect upon their lives or morals. Their villages are only found on the main river, hunting parties only going into the back country temporarily, at which time all the members of the families take part in the expedition. The population found on this part of the river is much larger than that of the Upper Yukon. There is no time of the year when more or less people are not to be found in the villages, and we find among them a larger proportion of females than on the Upper Yukon. Some time ago the lack of females was most noticeable among the Indians of the upper river, attributable to hard usage and the work they were compelled to do, as well as to the lack of care of female children. Of late, however, female children have been better taken care of, and probably in course of time there will be more marriageable women among them. Most of the married women to be seen there at present come from the Koyukuk or the Lower Yukon River. The Nuklukayet and Nowikaket people claim to have their origin from the tribe on the Koyukuk River in the north. The Tanana River and Upper Yukon Indians speak an entirely different language, though there is a dialect by which they can communicate with the various tribes.

The fur trade has undergone considerable change of late years, the catch of furs being considerable less than formerly, partly owing to the decrease of fur-bearing animals, and also to their being more white men in the country, independent of the fur traders, causing the circulation of more money among the natives, with which they buy instead of trading furs. The average catch of land furs for the whole year ranges from 16,000 to 20,000 pelts, usually with a large proportion of mink skins, the lowest-priced fur on the market.

There are six trading posts at points on the river in Alaska. The traders, to reach the back country, usually fit out trusty natives with small stock of goods to travel among the distant tribes. Since the discontinuance of opposition the white traders do not travel in the winter. The prices paid are regulated by the standard price of red fox or martin, called one skin, about $1.25. A prime beaver would be two skins, black bear four skins, lynx one skin, land otter two or three skins and so on. Five yards of drilling or one pound of tea or one pound of powder or half a pound of powder with one box of caps and one pound of shot are given for one skin, fifty pounds of flour for four skins, five pounds of sugar for one skin. These are sample prices obtained by the natives, with little variation, until the mining district is reached, where the prices are higher, to conform with the prices charged to miners.

The merchandise is carried on the river by

means of stern-wheel steamers, the two principle ones belonging to the Alaska Commercial Company, one of 200 tons, the other of thirty tons capacity, carrying freight and passengers. On the larger boat there is a white man for captain and another for engineer, but both captain and engineer are unlicensed and without papers; the rest of the crew are Indians. There are three other small steamers, two belonging to the Russian and Catholic missionaries respectively and one to the traders at Fort Selkirk. All supplies are received at St. Michael on Norton Sound, ninety miles north of the mouth of the Yukon, the furs and gold obtained being turned over to the Alaska Commercial Company's agent there and shipped to San Francisco. Once a year, in June, the missionaries and traders assemble at St. Michael's and for a few days that place is doing a rushing business. It has become a regular fair for the natives, who gather in numbers from various points on the coast and river, getting a few days' work from the company and having the satisfaction of seeing the new stock of merchandise.

The influx of miners to the country has produced marked changes among the natives, and not to their benefit morally. The illicit manufacture and use of liquor, both by the traders of the company and miners, is demoralizing the natives to a great extent. It is openly carried on both the upper and lower rivers. At Andreafsky, on the lower river, it is a common sight to see intoxicated natives, more especially in the winter, and the natives have now learned the process of making liquor themselves, more particularly on the coast and the Lower Yukon.

On the coast the temperature varies from 70 degrees Fahrenheit in summer to 40 degrees and 45 degrees below zero in the winter. The late summer and fall is usually stormy and wet, the snow fall in winter being from three to five feet on a level. Navigation is closed to the outside for seven months in the year by heavy ice on the sea. The Yukon River is closed by ice from November to the end of May. In the interior the climate is dryer and warmer in summer, but many degrees colder in winter, the thermometer going as low as 60 degrees below zero. The snow fall is excessive, but less wind prevails here in winter than on the coast.

For many miles on the lower river the banks are devoid of timber other than a stumpage growth of willow brush, alder and cottonwood. The first spruce timber is seen some fifty miles below the Russian mission, at Ikomiut, and from there up to its head the river is more or less belted with timber, spruce, fir, hemlock, birch, alder and cottonwood being the varieties most predominant.

On the low islands and flats the spruce attains a considerable size, but as lumber it is not adapted for any purpose beyond the needs of the miners and others in the country, being checked by frost and full of knots. The growth of timber seems to be entirely confined to the margins of the streams and rivers in many instances being merely a fringe on the banks.

There is a great variety of berries to be found all through the country; high and low bush cranberries, blueberries, salmon berries, red currants, and raspberries. The salmon or dewberries abound on the swampy lands of the Lower Yukon, and are gathered by the natives in quantities, who preserve them by burying them in the ground, using them as a delicacy in the winter, mixed with seal oil or deer fat and snow.

Game is said to be scarce, considering the immense stretches of uninhabited country. Numerous signs are to be seen on the banks of the main river, but so far few white men have proved successful hunters, owing to the difficulties of travel. An Indian traveling with no impediments can scour over the country, and, being acquainted with every game sign, can obtain some reward for his exertion, where a white man would starve.

Though some distance to the north of the entrance of the Yukon River, St. Michael has always been a controlling centre and basis of supplies for the great river of the far northwest. From here the hardy Muscovite pioneers pushed their advance slowly and laboriously with clumsy boats in skin-covered "bidars," and trudging over the frozen snow plains with their dog teams until the met the forerunners of the Hudson Bay Company on their way down the river, which English geographers of that time pictured as emptying into the Arctic.

As long as the Russians were in possession of this region all furs secured in the Kuskokwim Valley were transported over the Yukon portage to St. Michael, and thence shipped to Sitka, together with those obtained by barter from the natives of the shores and islands of Bering

Strait. The first American traders to engage in the Yukon trade were members of the Western Union Telegraph expedition, and foremost among these pioneers were Ketchum and Clark. Later came Mercier, a brother of the Canadian ex-minister, and a host of other French Canadians, together with three prospectors, McQueston, Mayo (Americans), and Harper (an Englishman), who still control the trade and much of the mining industry of the Upper Yukon and its tributaries from Fort Selkirk westward.

The basis of supplies for the whole district was early taken by the Alaska Commercial Company, who at first utilized a small stern-wheel steamer placed upon the river by the telegraph company, and later built other vessels for the purpose of towing loaded barges up the river. Later the firms who entered into competition with the company in other districts made a lodgment near St. Michael, and another steamer was placed upon the river.

In the year 1883 this opposition collapsed, but shortly after the bar diggings of Forty Mile Creek and other parts of the Upper Yukon were discovered, which caused a sudden revival of trade, chiefly in miners' supplies, and induced the traders mentioned above to acquire small steamboats of their own.

The flourishing missionary establishments of the Roman Catholic and the Episcopalian Churches also serve to increase traffic upon the great river during the brief season of navigation. Both the Roman Catholic and the Russian Orthodox missions now possess steamers for carrying their freight up from St. Michael and to transport their missionaries over their extensive field of labor.

The post of St. Michael, though insignificant in dimensions and most desolate surroundings, springs into life and activity once a year. With the first breath of spring, at the end of May, the up-river people shake off their winter's lethargy and prepare for their annual meeting with their fellows from the outside world. The steamers, which had been hauled up at various points on the river bank in the autumn, are prepared and launched once more upon the muddy waters as soon as the ice has ceased to float down the rapid current, crashing and grinding cake against cake, or pressing against the forest border of the channel, cutting and barking the trees, and down in the treeless waste of the lower river, undermining the soft clay banks and changing the face of the landscape.

Traders, missionaries, miners and natives crowd every craft and enjoy the hospitality freely offered them on their seaward progress at posts and missions.

By the end of June all these Yukon pilgrims have reached their goal, St. Michael, and, while they are waiting for the arrival of the ocean steamer, accounts are regulated and engagements entered into for the transactions for the coming season. The natives assembled here on these occasions represent all the tribes of the Yukon and many of those of the Arctic and the Bering Sea coast. Most of these bring trade with furs or ivory and whalebone, and, though all strive to hold their wares from the white man until the steamer arrives with the new stock of goods, quite an exchange of commodities goes on in the meantime among themselves.

With the arrival of the steamer, which is sometimes delayed weeks, causing much inconvenience to the commissary department of so large an assemblage, business activity rises at once to fever heat. Miners in ragged garments, showing the wear and tear of sub-arctic travel, Indians of the interior in beaded suits of tanned moose skin, and Eskimo in furs, all lend a hand and labor cheerfully, getting the cargo ashore and reloading it on the river boats. The black-robed missionary relaxes from his habitual dignity, and can be seen trundling barrels and bales and trucking boxes and miscellaneous packages over the planked walls of the crowded station. The light of day lasts all through the brief arctic summer night, and the turmoil is kept up almost without cessation until at last the steamer's whistle warns those who do not wish to spend another winter in these desolate regions that they must depart. The lucky individuals, who have bags of gold dust in the purser's safe, seek their comfortable staterooms, while the rank and file of prospectors cheerfully accept such accommodations as steerage or deck afford, brining out of the country no more, and probably much less, than they brought into it over the toilsome road from Chilkoot to the Yukon diggings.

The trade of the Upper Yukon is of great volume, but it is carried on under peculiar conditions. The supplies are purchased in the United States, chiefly in California, and carried thence to St. Michael. From here the river steamers,

THE WONDERFUL YUKON COUNTRY. 171

carrying the Stars and Stripes, ascend the river, dropping freight at intermediate stations, but the principal business is transacted at the point of junction between the Yukon River and Forty Mile Creek, some thiry miles beyond our boundary. The purchasers here are miners who toil in the upper ravines of Forty Mile Creek, which lie within the limits of Alaska. Prices are necessarily high, for during every winter the trader is called upon to feed numbers of unsuccessful miners and assist them in leaving the country in the spring.

The Alpine coast region, rising abruptly from the shores of the North Pacific Ocean between Cape Spencer on the east and Mount St. Elias on the west, has been the objective point of several exploring expeditions. Lieutenant Frederick Schwatka, formerly of the United States Army; Professor Libby and Lieutenant Selton-Karr, of the British Army, were among the first to attempt the exploration and partial ascent of Mt. St. Elias, a giant among the mountain peaks of North America.. They were followed later by well-organized parties under the auspices of the National Geographic Society and the United States Geological Survey. Under the leadership of

Professor I. C. Russell these parties have attained in two successive seasons a large amount of the most valuable information concerning this mountain, which is claimed by Americans and English alike as lying within their boundaries.

In the course of his second exploration Professor Russell, after reaching a height of 14,000 feet, succeeded in making measurements of Mt. St. Elias from a base line on the sea shore, from which the height of the mountain was computed, at 18,100 feet. On the return journey the low coast region lying at the foot of the observations was made in Disecnhantment Bay, at the head of Yakutat Bay, furnishing material for the compilation of a reliable map of the estuary, exhibiting a remarkable deviation from the outlines heretofore accepted on the authority of Tebenkof and others, who did not personally explore the innermost recesses of this great bay.

Another important exploration, resulting in the collection of much information concerning the interior geography and topography in Alaska and adjoining territory in the British possessions was made by Lieutenant Frederick Schwatka, accompanied by Lieutenant Hayes, of the United States Geological Survey. This expedition set out in

THE WONDERFUL YUKON COUNTRY. 173

an easterly direction from Taku Inlet along Taku river; then crossing the coast range, they emerged upon the banks of Lake Aklene, which is probably the true head of the Yukon River. Following the northern outlet of this lake, the party passed the mouth of the tributary heretofore accepted as the Yukon's head, a few miles above Lake Labarge. Thence to Fort Selkirk their course was over a well-known course, but on leaving that point an entirely new route was followed, leading towards the mountains forming the divide between the Yukon Basin, the upper course of White River, and the easternmost tributary of Copper River. After discovering a pass but little over 5000 feet in height the party struck the Chityna River, about midway between its headwaters and its junction with the Copper. The latter river was then followed to the coast.

Valuable additions have also been made to our knowledge of Alaskan geography by the members of an exploring expedition organized in 1890 under the auspices of Frank Leslie's Illustrated Weekly. The leaders of the party, Messrs. A. J. Wells, E. J. Glave and A. B. Schanz, entered the interior by way of the Chilkoot River, and, after crossing the coast range, came upon a large lake, the head of the Tahkina tributary of the Yukon, which was named Lake Arkell. It is probable that this is the same lake which the German explorer, Krause, visited in 1879 and named Western Kussoa, in contra-distinction from the Eastern Kussoa which he found beyond the Chilkoot Pass. Here Mr. Glave left the party, and, striking across the coast range southward, discovered the headwaters of the Alsekh River; following down its channel to the coast at Dry Bay, Messrs. Wells and Schanz proceeded to the Upper Yukon by the usual route. At Forty Mile Creek Mr. Wells and another white man turned off, and, with the assistance of a miner who was engaged as guide, crossed over into the basin of the Tanana River and explored an unknown tributary of that stream. Mr. Schanz traveled down the Yukon to St. Michael, and thence back to the Kuskokwim Portage and down that river to the sea coast, reaching Bristol Bay in October. Here he was joined a month loter by Mr. Wells and his party, who had followed the same route from the mouth of the Tanana River. During the months of January and February Mr. Schanz, in company with Mr. J. W. Clark, accomplished a dog sledge journey of discovery, resulting in the definite location and exploration of a large lake to the northward of Lake Lliamna. This important sheet of water, some seventy-five miles long, was named Lake Clark. The Noghelin River, broken about midway by a magnificent fall, connects it with Lake Lliamna, of which it is the principal feeder.

CHAPTER X.

THE BOUNDARY DISPUTE.

There is every probability that the new discoveries of gold will bring the long pending boundary dispute between the United States and Great Britain to a head, for the most profitable diggings are situated near the 141st meridian, which is the boundary fixed by treaty in the Northwestern territory. Unfortunately this country has never been adequately surveyed, and there is great uncertainty as to whether some of the richest creeks are situated on Alaskan or British soil. The Canadian authorities have lost no time in assuming that the boundary line should be so drawn as to bring all the finest deposits to the East. There is no doubt that Klondike River is in Canadian territory, but the Canadian newspapers claim persistently that Miller's Creek and the other gold yielding creeks near by are east of the line.

It appears from the report of General W. W. Duffield, Chief of the United States Coast and Geodetic Survey, that in spite of Klondike, the United States has the lion's share of the gold fields.

"In the Yukon region," he says, "the surveys of the representatives of the United States and those of Great Britain are very nearly identical, with one or two exceptions, and are remarkable when all things are considered. Her Majesty's surveyor, Ogilvie, was appointed by Mr. King, and went up there in 1890. The coast survey tested the work of Ogilvie and the Canadians on Forty-Mile Creek and on the Yukon. They found at Forty-Mile Creek a pine tree marked by Ogilvie, which lacked only fifteen-one-hundredths of a second of being identical with the determination of the Coast Survey. In that latitude this makes a difference of six feet and seven inches. On the Yukon, Ogilvie marked a willow tree on the south or left bank and a pine on

THE BOUNDARY DISPUTE.   177

the right or north bank. When these were tested they were found to be fourteen seconds and 22-100ths out. This in that latitude is a distance of 618 feet. One of these marks—that at Forty-Mile Creek—is too far west, so that the United States loses six feet and seven inches. On the Yukon the point is too far east, so that the United States gains and Canada loses 618 feet. The Coast Survey has marked the crossing of the 141st meridian at Porcupine Creek, but the Canadians have never tested it. Considering the fact that Ogilvie was traveling light on snow shoes, and that almost all of his determinations were made with the sextant, his work is excellent.

"In substance, these determinations throw the diggings at the mouth of Forty-Mile Creek within the territory of the United States. The whole valley of Birch Creek, another most valuable gold producing part of the country, is also in the territory of the United States. Most of the gold is to the west of the crossing of the 141st meridian at Forty-Mile Creek.

"If we produce the 141st meridian on a chart, the mouth of Miller's Creek, a tributary of Sixty-Mile Creek, and a valuable gold region, is five miles west in an air line, or seven miles according to the windings of the stream—all within the territory of the United States. In substance, the only places in the Yukon region where gold in quantities has been found, are, therefore, all to the west of the boundary line between Canada and the United States."

In spite of this the very latest Canadian map claims Miller's Creek and Glacier Creek.

This is not all. For the last twelve years the British Government has been trying by continually increasing claims to shake the hold of the United States upon the strip of mainland in Southeastern Alaska, and upon some of the gold bearing islands.

Up to 1884 both countries were practically at one as to the boundary line from Mt. St. Elias to the southeast. According to the terms of the treaty between Russia and Great Britain, the United States, in purchasing Alaska in 1867, acquired all of Russia's rights. In describing the southeastern boundary the Anglo-Russian treaty reads:

"The line of demarkation between the possessions of the high contracting parties upon the coast of the continent and the islands of America

A FAR LOOK TOWARD THE GOLDFIELDS.

THE BOUNDARY DISPUTE. 179

to the northwest shall be drawn in the following manner: Commencing from the southernmost point of the island called Prince of Wales Island, which point lies in the parallel of 54 degrees 40 minutes north latitude, and between the 131st degree and the 133d degree of west longitude, the same line shall ascend to the north along the channel called Portland Channel, as far as the point of the continent where it strikes the 56th degree of north latitude; from this last-mentioned point the line of demarkation shall follow the summit of the mountains situated parallel to the coast as far as the point of intersection of the 141st degree of west longitude (of the same meridian), and finally, from the said point of intersection, the said meridian line of the 141st degree, in its prolongation as far as the frozen ocean, shall form the limit between the Russian and British possessions on the continent of America to the northwest.

"Whenever the summit of the mountains which extend in a direction parallel to the coast from the 56th degree of north latitude to the point of intersection of the 141st degree of west longitude, shall prove to be at the distance of more than ten marine leagues from the ocean, the limit between the British possessions and the line of coast which is to belong to Russia, as above mentioned (that is to say, the limit to the possessions ceded by this convention), shall be formed by a line parallel to the winding of the coast, and which shall never exceed the distance of ten marine leagues therefrom."

On all maps from 1825 down to 1884 the boundary line, it was declared, had been shown as, in general terms, parallel to the winding of the coast, and thirty-five miles from it. In 1884, however, an official Canadian map showed a marked deflection in this line at its south end. Instead of passing up Portland Canal this Canadian map showed the boundary as passing up Behm Canal, an arm of the sea some sixty or seventy miles west of Portland Canal, this change having been made on the bare assertion that the words Portland Canal as inserted were an error. By this change the line and an area of American territory about equal in size to the State of Connecticut was transferred to British territory. There are several facts which militate against this claim. In the first place, the British Admirality, when surveying the northern limit of British Columbian possession, in 1868, one year after the cession of Alaska, surveyed Portland Canal, and not Behm Canal, thus by implication admitting this canal as the boundary line. The region now claimed by British Columbia was at that time occupied by a military post of the United States without objection or protest on the part of British Columbia. Annette Island, in this region, was, by an act of Congress, four years ago, set apart as a reservation for the use of the Metlakatla Indians. Within a year the United States Engineers, by authorization of Congress, have made an official survey of the west bank of Portland Canal, building stone houses at various places, and thus exercising an undoubted act of sovereignty.

Another grab was made at Lynn Canal, the northernmost extension of the Alexander Archipelago, which runs north of Juneau, and is the land outlet for the Yukon trade. The official Canadian map of 1884 carried the boundary line around the head of this canal; another Canadian canal map three years later carried the line across the head of the canal in such manner as to throw its head-waters into British territory; still later, Canadian maps carry the line not across the head of the canal, but cross near its mouth, some sixty or seventy miles south of the former line, in such a way as to practically take in Juneau, or, at least, all the land immediately back of it. And the very latest official map, just published at Ottawa, while it runs no line southeast of Alaska, prints the legend "British Columbia" over portions of the Lynn Canal that are now administered by the United States. The post of Dyea, marked Ty-a or Tyea, on Canadian maps, which is at the head of navigation on Lynn Canal, and where the trail starts into the interior for the Yukon, which was made a sub-port of entry the other day by Secretary Gage, is claimed by the Canadians.

CHAPTER XI.

GOLD PRODUCTION OF THE WORLD.

The United States is the chief gold producing country in the world. We have held the lead ever since the discovery of gold in California, with the exception of 1894, when we fell to third place, surrendering first place to Australia and

FORT YUKON, ON THE ARCTIC CIRCLE.

the second to Africa. The United States regained in 1895 the place lost in 1894, its output of gold in the former year having exceeded that of 1894 by $7,110,000. In 1895 the gold yield of the United States was 2,254,760 ounces fine, valued at $46,610,000, while the yield of Australasia was 2,167,117 ounces, valued at $43,893,300. The latest reported findings, as it happens, will not properly be credited to the United States, but with the development of the fields actually situated on the Alaskan side of the boundary, there is almost sure to be such an addition to the product of the United States mines as to place them easily and permanently at the head of the gold-producing countries of the world. The largest production of gold in the United States for any single year was $65,000,000, in 1853. The next most productive years were 1852 and 1854, when the returns were $60,000,000 for each year. The least productive year since the gold discoveries in California was 1883, when only $30,000,000 was mined. Since then the advance has been steady.

The gold output of the world from the time of the discovery of America to the close of the fiscal year 1895 has been estimated at $8,781,858,-700. The California gold field since their discovery in 1849 have alone yielded $2,035,416,000. The total output of the Australian mines, which were first worked in 1851, has been $1,655,713,-000, and gold has been taken out of the mines of South Africa since 1890 to the amount of $211,-632,990. The total production of the world in the last three years has been as follows:

1893—$157,494,800.
1894—$181,567,800.
1895—$200,215,700.

Of these amounts considerably more than 50 per cent. has come from the United States, Australasia, Russia and South Africa, as follows:

	1893. Dollars.	1894. Dollars.	1895. Dollars.
United States	35,955,000	39,500,000	46,610,000
Australasia	35,688,600	41,760,800	44,798,300
Russia	27,808,200	24,133,400	28,894,400
Dominion of Canada	927,200	1,042,100	1,910,900
Africa	28,943,500	40,271,000	44,554,900

CHAPTER XII.

CHARACTERISTICS OF OUR NORTHWESTERN POSSESSIONS.

By JOHN F. PRATT.

To those who are familiar with the story of the northwestern country the rich discoveries of gold in the Yukon Valley are no surprise. They form a chapter in the gold findings of that region which has been writing for many years. Just before the war there was widespread excitement over the discovery of gold in the Caribou district of British Columbia, and the diggings there for a time were very rich. The craze resulted in much hardship and many deaths. Later, subsequent to the purchase of Alaska, gold was found in considerable quantities in the Cassiar district, farther to the northwest in British America.

The Cassiar Mountains are situated between the 60th and 65th degrees of north latitude, at the headwaters of the Pelly River. They are reached by way of the Stikine River, the outlet of which is near Fort Wrangell. These diggings are still carried on, and they have yielded much gold. There are several quartz lodes in the Cassiar district which are rich, but hardly rich enough to mine profitably with the present inadequate facilities for reaching them and for transporting machinery. During high water steamboats can run well up the river, leaving a distance of forty miles between Telegraph Creek and Moose Lake to be traveled by pack trains. The Cassiar diggings are far less accessible than the new gold fields.

Now, in this same trend or general direction, as if in continuation of the line running northwest from Caribou through the Cassiar range, come the diggings near the place where the Yukon River crosses the boundary between Alaska and British North America, and we are bound to suppose that the lode runs still farther along toward the northwest into the country which is not yet prospected at all.

Although the Klondike is on the Canadian side of the boundary, there is reason to believe that the great bulk of the gold territory is west of the boundary on the American side. This is to be deduced from the peculiar locations of the streams from along which gold has thus far been taken. Sixty Mile Creek and Forty Mile Creek lie largely in United States territory. Both flow into the Yukon toward the east. Birch Creek flows into the Yukon toward the north and the Tenanah River toward the northwest. The Sushitna flows toward the south into Cook's Inlet, on the southern coast, where gold has been found. The headwaters of all these gold-bearing streams flowing in different directions are thus seen to be in the same country, about 100 miles west of the boundary and south of the Yukon. This seems to indicate that the great mother lode is probably within the United States and that the more permanent diggings will be found in United States territory centering about a spot not 100 miles west of the boundary.

The diggings around Klondike, therefore, are not in the middle of the richest gold territory, but are rather on the northeast edge. Gold has been found as far west as Cook's Inlet on the southern coast, between the 150th and 152d degrees west longitude, and it has been found as far east as the 128th degree. There is a gold-bearing area of between forty thousand and fifty thousand square miles, and the best part of it is on the United States side of the boundary.

Of course, our actual information is exceedingly limited. Perhaps we know less about the Alaskan Territory than about any other territory of equal size on the continent. The Yukon River has been explored from its mouth to the region of the gold diggings, and the trail of the miners from Chilkoot Pass to the diggings has given us a knowledge of that region; we know the mouths of the streams as they flow into the Yukon; but aside from these we have learned very little. Travelers and prospectors have found out more or less by interviewing the Indians, who have a general idea of direction and distance, but this knowledge is not exact. Even our most elaborate maps of Alaska depend upon miners' plottings and not upon official surveys for the location of the creeks and rivers in the gold region. What other information we have of the interior has been acquired largely from prospectors and on the British side of the boundary from Canadian explorers. We know something about the streams and the outlets, but we have not discovered their sources. The hill country is practically unknown, and there may be large streams concerning which we have no information. There is an immense stretch of territory of perhaps 250,-

000 square miles of which we are practically ignorant.

We are about as badly off with regard to the coast line. The southern coast we know fairly well in a general way, but there has never been an official survey beyond Sitka. Even the maps of the Aleutian Islands are inherited from Russia, and there has never been anything like a survey of the mouth of the Yukon River. It would be of great value, now, if we knew whether there was a channel through which the Yukon could be reached from Bering Sea by deep water ships. We are aware now that shoals extend out for twenty-five miles, apparently stretching all the way across the mouth of the Yukon, but there has never been any survey to discover whether there might not be a passage through. All ships now, owing to lack of knowledge concerning these shoals, are compelled to avoid them altogether by going to St. Michael, thirty miles north of the mouth of the river, there to meet the river boats which are obliged, on that account, to make the dangerous trip outside on the ocean. If seagoing ships could be brought into the mouth of the Yukon they might proceed up the river at least to as great a distance as that between New

Orleans and the mouth of the Mississippi and possibly they could continue the journey for several hundred miles. There should be an early appropriation for a survey of the mouth of the Yukon. There should also be a survey to discover whether some of the portages between the Yukon at different points and Bering Sea might not be available for general traffic. At one place not far from St. Michael Island the Yukon in its windings approaches within a few miles of the coast.

It is peculiar that the two entrances to the gold country should be, one through the head of the Yukon River, the other through the mouth. It is 2000 miles from Sitka to the mouth of the Yukon, and from the mouth of the Yukon to the Klondike is about the same distance, for the river is very winding throughout its course. The route by sea, which takes the traveler through Unimak Pass, separating two of the Aleutian Islands, to St. Michael, and thence by river boat to his destination, will be used largely for getting supplies into the gold country; but it is a long journey, and steamers going up the Yukon have to wait until the ice leaves the river.

For miners the trails leading up from the head-

LAKE BENNETT FROM THE SOUTH.

waters of Lynn Canal will be more convenient. There are four entrances into the gold country from the coast in this direction, one by the Taku River, just below Juneau, the others by the White Pass, the Chilkoot Pass and the Chilkat Pass. Of these only two, the Chilkoot Pass and the White Pass, are really feasible, and the Chilkoot Pass is so much the better of the two that it is the route almost exclusively used. It is superior to the others because it has a shorter distance to travel and is not so rugged. White Pass is rugged throughout its entire length, and over a great part of it travelers would be compelled to go on their hands and knees. The route by the Taku River is very rough and requires many miles of packing. The Chilkat route is what is known as Jack Dalton's trail. Dalton is well known in all that country. He was a scout for Glaves, who was the first white man to explore the region, and for the last few summers he has been engaged in carrying whisky and various supplies up into the mining camps. He fits out at Juneau, with his nine horses, a white man and two Indians, crosses to the Chilkat Inlet, and then strikes off into the wilderness toward the headwaters of the White River. Nobody else knows exactly what route he takes, and he will not tell.

Lynn Canal, as it approaches its head, divides into two branches, Chilkat Inlet on the west and Chilkoot Inlet on the east. Chilkoot Inlet in turn has a branch known as Dyea Inlet, and at the head of Dyea Inlet is a small Indian village and a store known as Healey's store. In 1894 Healey's store was the only house in the place. It acquires its importance because it is the head of navigation and the last base of supplies for miners before striking off into the trail for the gold country.

If a railroad is ever constructed into the gold fields it will probably be through Chilkoot Pass.

The natives of the gold country in the interior are known as Stick Indians. "Stick" is the Chinook expression for wood, and the Stick Indian consequently is the Indian of the interior or forest. He is quite distinct from the Chilkat Indian on the coast. He is short of stature, but stout, his diminutiveness being due to the hardships and privations which he has been compelled to suffer always. But physically he is very strong. He can carry on his back all day a pack which many men would find it uncomfortable to lift.

There are marked differences between the Chilkats and the Sticks. The Chilkats spend most of their time on the streams and use canoes almost exclusively. They do very little tramping. They are a fine race, hardy and well formed. The Sticks never use canoes. Some of them have little dug-outs in the streams in their own country, but when they come down to the coast, as they come occasionally now, they are quite lost.

The Stick Indians are centered around the streams of Alaska, and have to keep pretty near to the main stream in order to get their food. Until very recently they have never dared to come down below the Chilkat Pass, so completely were they terrorized by the Chilkat Indians. The Chilkats have had absolute control of the country along the coast, so much so that they were able to collect toll from the miners who first went through the Chilkoot Pass. When the Russians were in possession of Chilkoot the Chilkats were a kind of middlemen between the Russian traders and the Indians of the interior. Indeed, these peculiar relations seem to have had a great deal to do with the drawing of the boundary line between British America and Russian

America. The idea of Russia was to continue the line of demarkation between the trading settlements of the coast and the Indian settlements of the interior, so that this line is really not a geographical line, but is intended rather to mark the extent of the control of the Indians of the coast; that is, to the summit of the mountain ranges extending from Portland Canal north to Mount St. Elias, beyond which the Sticks never dared to come.

CHAPTER XIII.

LAWS GOVERNING THE LOCATION OF CLAIMS.

It is important to know something about the laws of the United States and of Canada which govern the patenting of mineral lands and which must be observed in locating claims. The public land laws of the United States do not apply to Alaska, and neither do the coal land regulations, which are distinct from the mineral regulations. The Territory of Alaska is expressly excluded from the operations of the public land and coal land laws by provisions of the laws themselves. Mineral lands have been patented in Alaska since 1884. Hon. Binger Hermann, Commissioner of the United States General Land Office, has authorized the statement that the following laws are applicable to the Territory:

First—The mineral land laws of the United States.

Second—Town-site laws, which provide for the incorporation of town sites and acquirement of title thereto from the United States Government by the town-site trustees.

Third—The laws providing for trade and manufactures, giving each qualified person 160 acres of land in a square and compact form.

The act approved May 17, 1884, providing a civil government for Alaska, has this language as to mines and mining privileges:

"The laws of the United States relating to mining claims and rights incidental thereto shall, on and after the passage of this act, be in full force and effect in said district of Alaska, subject to such regulations as may be made by the Secretary of the Interior and approved by the President."

"Parties who have located mines or mining privileges therein, under the United States laws applicable to the public domain, or have occupied or improved or exercised acts of ownership over such claims, shall not be disturbed therein, but shall be allowed to perfect title by payment so provided for."

There is still more general authority. Without the special authority, the act of July 4, 1866, says: "All valuable mineral deposits in lands belonging to the United States, both surveyed and unsurveyed, are hereby declared to be free and open to exploration and purchase, and lands in which these are found to occupation and purchase by citizens of the United States and by those who have declared an intention to become such, under the rules prescribed by law and according to local customs or rules of miners in the several mining districts, so far as the same are applicable and not inconsistent with the laws of the United States."

Under United States laws only those who are citizens or who have declared intention to become citizens may locate or buy claims. There are no "free miners." The government cannot give the right to mine except in public lands, and these must contain valuable mineral deposits. A claim may not exceed beyond 1500 feet along a vein or 300 feet on each side of the middle of the vein. A person may locate a claim through an agent; $100 worth of work must be done each year. Local government prevails in the various mining districts of the United States, each district being free to manage its own affairs so long as it does not do anything inconsistent with the national laws.

Mining operations on the Klondike on the British side of the boundary are subject, not to the regulations of the Province of British Columbia, but to the general mining laws of the Dominion of Canada.

As soon as the mounted police force has been raised to 100 men from the 20 men now keeping order in the country, it will be considered safe to promulgate the new regulations for placer gold mining. These provide that every alternate claim is to be reserved by the crown for the public benefit, and that the royalty to the crown is to be 10 per cent. on the yield up to $500 a month and 20 per cent. over $500 a month.

A difficulty with respect to the alternate claims is that the placer territory is already staked solid by prospectors, so far as they have gone. Turning out the prospectors on every alternate claim is not likely to prove a pleasant proceeding. Many have staked without registering, and those only who have registered are safe. The registering has been raised from $5 to $15 in each case, with an annual tax of $100.

The miners' tax applies to all alike, and will not be levied so as to discriminate against Americans. It will be almost impossible to collect more than a small proportion. As regards its effects on the Canadian miners, it will undoubtedly drive the majority of them, as soon as they have made their pile, to take it to the United States to evade full assessment. How much of the royalties will ever find their way to Canada is a question.

A digest of the Dominion mining laws is given below:

PLACER MINING.

Nature and Size of Claims—For "Bar Diggings:" A strip of land 100 feet wide at high-water mark and thence extending into the river at its lowest water level.

For "Dry Diggings:" One hundred feet square.

LAWS GOVERNING LOCATION OF CLAIMS.

For "Creek and River Claims:" Five hundred feet along the direction of the stream, extending in width from base to base of the hill or bench on either side. The width of such claims, however, is limited to 600 feet when the benches are a greater distance apart than that. In such a case claims are laid out in areas of ten acres with boundaries running north and south, east and west.

For "Bench Claims:" One hundred feet square.

Size of claims to discoverers or parties of discoverers:

To one discoverer, 300 feet in length; to a party of two, 600 feet in length; to a party of three 800 feet in length; to a party of four, 1000 feet in length; to a party of more than four, ordinary sized claim only.

New strata of auriferous gravel in a locality where claims are abandoned, or dry diggings discovered in the vicinity of bar diggings, or vice versa, shall be deemed new mines.

Rights and Duties of Miners—Entries of grants for placer mining must be renewed and entry fee be paid every year.

No miner shall receive more than one claim

in the same locality, but may hold any number of claims by purchase, and any number of miners may unite to work their claims in common, provided an agreement be duly registered and a registration fee of $5 be duly paid therefor.

Claims may be mortgaged or disposed of provided such disposal be registered and a registration fee of $2 be paid therefor.

Although miners shall have exclusive right of entry upon their claims for the "miner-like" working of them, holders of adjacent claims shall be granted such right of entry thereon as may seem reasonable to the superintendent of mines.

Each miner shall be entitled to so much of the water not previously appropriated flowing through or past his claim as the superintendent of mines shall deem necessary to work it, and shall be entitled to drain his own claim free of charge.

Claims remaining unworked on working days for seventy-two hours are deemed abandoned, unless sickness or other reasonable cause is shown or unless the grantee is absent on leave.

For the convenience of miners on back claims, on benches or slopes, permission may be granted

LAWS GOVERNING LOCATION OF CLAIMS. 201

by the superintendent of mines to tunnel through claims fronting on water courses.

In case of death of a miner the provisions of abandonment do not apply during his last illness after his decease.

Acquisition of Mining Locations—Marking of Locations: Wooden posts, four inches square, driven eighteen inches into the ground and projecting eighteen inches above, must mark the four corners of a location. In rocky ground, stone mounds three feet in diameter may be piled about the post. In timbered land well-blazed lines must join the posts. In rolling or uneven localities, flattened posts must be placed at intervals along the lines to mark them, so that subsequent explorers shall have no trouble in tracing such lines.

When locations are bounded by lines running north and south, east and west, the stake at the northeast corner shall be marked by a cutting instrument or by colored chalk, "M. L., No. 1" (mining location, stake No. 1.) Likewise the southeasterly stake shall be marked "M. L., No. 2," the southwesterly "M. L., No. 3," and the northwesterly "M. L., No. 4." Where the boundary lines do not run north and south, east and west, the northerly stake shall be marked 1, the easterly 2, the southerly 3 and the westerly 4. On each post shall be marked also the claimant's initials and the distance to the next post.

Application and Affidavit of Discoverer: Within sixty days after marking his location, the claimant shall file in the office of the Dominion Land office for the district a formal declaration, sworn to before the land agent, describing as nearly as may be the locality and dimensions of the location. With such declaration he must pay the agent an entry fee of $5.

Receipt Issued to Discoverer: Upon such payment the agent shall grant a receipt authorizing the claimant, or his legal representative, to enter into possession, subject to renewal every year, for five years, provided that in these five years $100 shall be expended on the claim in actual mining operations. A detailed statement of such expenditure must also be filed with the agent of the Dominion lands, in the form of an affidavit corroborated by two reliable and disinterested witnesses.

Annual Renewal of Location Certificate: Upon payment of the $5 fee therefor, a receipt shall be issued entitling the claimant to hold the location for another year.

Working in Partnership: Any party of four or less neighboring miners, within three months after entering, may, upon being authorized by the agent, make upon any one of such locations, during the first and second years, but not subsequently the expenditure otherwise required on each of the locations. An agreement, however, accompanied by a fee of $5, must be filed with the agent. Provided, however, that the expenditure made upon any one location shall not be applicable in any manner or for any purpose to any other location.

Purchase of Location: At any time before the expiration of five years from date of entry a claimant may purchase a location upon filing with the agent proof that he has expended $500 in actual mining operations on the claim and complied with all other prescribed regulations. The price of a mining location shall be $5 per acre, cash.

On making an application to purchase, the claimant must deposit with the agent $50, to be deemed as payment to the government for the survey of his location. On receipt of plans and field notes and approval by the Surveyor-General, a patent shall issue to the claimant.

Revision of Title: Failure of a claimant to prove within each year the expenditure prescribed, or failure to pay the agent the full cash price, shall cause the claimant's right to lapse and the location to revert to the crown, along with the improvements upon it.

Rival Claimants: When two or more persons claim the same location the right to acquire it shall be in him who can prove he was the first to discover the mineral deposit, and to take possession in the prescribed manner. Priority of discovery alone, shall not give the right to acquire. A subsequent discoverer, who has complied with other prescribed conditions, shall take precedence over a prior discoverer who has failed so to comply.

When a claimant has, in bad faith, used the prior discovery of another and has fraudulently affirmed that he made independent discovery and demarcation, he shall, apart from other legal consequences, have no claim, forfeit his deposit and be absolutely debarred from obtaining another location.

Rival Applicants: Where there are two or more applicants for a mining location, neither of whom is the original discoverer, the Minister of the Interior may invite competitive tenders or put it up for auction, as he sees fit.

Transfer of Mining Rights—Assignment of Right to Purchase: An assignment of the right to purchase a location shall be indorsed on the back of the receipt or certificate of assignment, and execution thereof witnessed by two disinterested witnesses. Upon the deposit of such receipt in the office of the land agent, accompanied by a registration fee of $2, the agent shall give the assignee a certificate entitling him to all the rights of the original discoverer. By complying with the prescribed regulations such assignee becomes entitled to purchase the location.

QUARTZ MINING.

Regulations in respect to placer mining, so far as they relate to entries, entry fees, assignments, marking of locations, agents' receipts, etc., except where otherwise provided, apply also to quartz mining.

Nature and Size of Claims—A location shall not exceed the following dimensions: Length, 1500 feet; breadth, 600 feet. The surface boundaries shall be from straight parallel lines, and its boundaries beneath the surface the planes of these lines.

Limit of Number of Locations—Not more than one mining location shall be granted to any one individual claimant upon the same lode or vein.

Mill Sites—Land used for milling purposes may be applied for and patented, either in connection with or separate from a mining location, and may be held in addition to a mining location, provided such additional land shall in no case exceed five acres.

GENERAL PROVISIONS.

Decision of Disputes—The superintendent of mines shall have power to hear and determine all disputes in regard to mining property arising within his district, subject to appeal by either of the parties to the Commisisoner of Dominion Lands.

Leave of Absence—Each holder of a mining location shall be entitled to be absent and suspend work on his diggings during the "close" season, which "close" season shall be declared by the agent in each district, under instructions from the Minister of the Interior.

The agent may grant a leave of absence pending the decision of any dispute before him.

Any miner is entitled to a year's leave of absence upon proving expenditure of not less than $200 without any reasonable return of gold.

The time occupied by a locator in going to and returning from the office of the agent or of the superintendent of mines shall not count against him.

Additional Locations—The Minister of the Interior may grant to a person actually developing a location an adjoining location equal in size, provided it be shown to the Minister's satisfaction that the vein worked will probably extend beyond the boundaries of the original location.
Forfeiture—In event of the breach of the regulations, a right or grant shall be absolutely forfeited, and the offending party shall be incapable of subsequently acquiring similar rights, except by special permission of the Minister of the Interior.

CHAPTER XIV.

CLIMATE OF ALASKA.

Willis L. Moore, chief of the United States Weather Bureau, has prepared a valuable and interesting report on the climate of Alaska. "The climates of the coast and the interior," he says, "are unlike in many respects, and the differences are intensified in this, as perhaps in few other countries, by exceptional physical conditions. The natural contrast between land and sea is here tremendously increased by the current of warm water that impinges on the coast of British Columbia, one branch flowing northward toward Sitka, and thence westward to the Kadiak and Shumagin Islands.

"The fringe of islands that separates the mainland from the Pacific Ocean from Dixon Sound northward and also a strip of the mainland for possibly twenty miles back from the sea, following the sweep of the coast, as it curves to the northwestward to the western extremity of Alaska, form a distinct climate division, which may be termed temperate Alaska. The temperature rarely falls to zero; winter does not set in until December 1, and by the last of May the snow has disappeared except on the mountains. The mean winter temperature of Sitka is 32.5, but little less than that of Washington, D. C. While Sitka is fully exposed to the sea influence, places further inland, but not over the coast range of mountains, as Killisnoo and Juneau, have also mild temperatures throughout the winter months. The temperature changes from month to month in temperate Alaska are small, not exceeding twenty-five degrees from midwinter to midsummer. The average temperature of July, the warmest month of summer, rarely reaches 55 degrees, and the highest temperature of a single day seldom reaches 75 degrees.

"The rainfall of Temperate Alaska is notorious the world over, not only as regards the quantity that falls, but also as to the manner of its falling, viz., in long and incessant rains and drizzles. Cloud and fog naturally abound, there being on an average but sixty-six clear days in the year.

"Alaska is a land of striking contrasts, both in climate as well as topography. When the sun shines the atmosphere is remarkably clear, the scenic effects are magnificent; all nature seems to be in holiday attire. But the scene may change very quickly; the sky becomes overcast; the winds increase in force; rain begins to fall; the evergreens sigh ominously, and utter desolation and loneliness prevail.

"North of the Aleutian Islands the coast climate becomes more rigorous in winter, but in summer the difference is much less marked. Thus, at St. Michael, a short distance north of the mouth of the Yukon, the mean summer temperature is 50 degrees, but four degrees cooler than Sitka. The mean summer temperature of Point Barrow, the most northerly point in the United States, is 36.8 degrees, but four-tenths of a degree less than the temperature of the air flowing across the summit of Pike's Peak, Col.

"The rainfall of the coast region north of the Yukon delta is small, diminishing to less than ten inches within the arctic circle.

"The climate of the interior, including in that designation practically all of the country except a narrow fringe of coastal margin and the territory before referred to as temperate Alaska, is one of extreme rigor in winter, with a brief, but

CLIMATE OF ALASKA.   211

relatively hot, summer, especially when the sky is free from clouds.

"In the Klondike region in midwinter the sun rises from 9.30 to 10 A. M., and sets from 2 to 3 P. M., the total length of daylight being about four hours. Remembering that the sun rises but a few degrees above the horizon, and that it is wholly obscured on a great many days, the character of the winter months may easily be imagined.

"We are indebted to the United States Coast and Geodetic Survey for a series of six months' observations on the Yukon, not far from the site of the present gold discoveries. The observations were made with standard instruments, and are wholly reliable. The mean temperature of the months October, 1889, to April, 1890, both inclusive, are as follows: October, 33 degrees; November, 8 degrees; December, 11 degrees below zero; January, 17 degrees below zero;, February, 15 degrees below zero; March 6 degrees above zero; April 20 degrees above. The daily mean temperature fell and remained below the freezing point (32), from November 4, 1889, to April 21, 1890, thus giving 168 days as the length of the closed season of 1889-90, assuming the outdoor operations are controlled by temperature only.

The lowest temperature registered during the winter were: 32 degrees below zero in November, 47 below in December, 59 below in January, 55 below in February, 45 below in March, 26 below in April.

"The greatest continuous cold occurred in February, 1890, when the daily mean for five consecutive days was 47 degrees below zero. The weather moderated slightly about the 1st of March, but the temperature still remained below the freezing point. Generally cloudy weather prevailed, there being but three consecutive days in any month with clear weather during the whole winter. Snow fell on about one-third of the days in winter, and a less number in the early spring and late fall months.

"Greater cold than that here noted has been experienced in the United States for a very short time, but never has it continued so very cold for so long a time. In the interior of Alaska the winter sets in as early as September, when snow storms may be expected in the mountains and passes. Headway during one of these storms is impossible, and the traveler who is overtaken by one of them is indeed fortunate if he escapes with his life. Snow storms of great severity may occur in any month from September to May, inclusive.

"The changes of temperature from winter to summer are rapid, owing to the great increase in the length of the day. In May the sun rises at about 3 A. M. and sets about 9 P. M. In June it rises about 1.30 in the morning and sets at 10.30, giving about twenty hours of daylight and diffused twilight the remainder of the time.

"The mean summer temperature of the interior doubtless ranges between 60 and 70 degrees, according to elevation, being highest in the middle and lower Yukon Valleys."

HENRY ALTEMUS' PUBLICATIONS.
PHILADELPHIA, PA.

STEPHEN. A SOLDIER OF THE CROSS, by Florence Morse Kingsley, author of "Titus, a Comrade of the Cross." "Since Ben-Hur no story has so vividly portrayed the times of Christ."—*The Bookseller*. Cloth, 12mo., 369 pages. $1.25.

PAUL. A HERALD OF THE CROSS, by Florence Morse Kingsley. "A vivid and picturesque narrative of the life and times of the great Apostle." Cloth, ornamental, 12mo., 450 pages, $1.50

VIC. THE AUTOBIOGRAPHY OF A FOX TERRIER, by Marie More Marsh. "A fitting companion to that other wonderful book, 'Black Beauty.'" Cloth, 12mo., 50 cents.

WOMAN'S WORK IN THE HOME, by Archdeacon Farrar. Cloth, small 18mo., 50 cents.

THE APOCRYPHAL BOOKS OF THE NEW TESTAMENT, being the gospels and epistles used by the followers of Christ in the first three centuries after his death, and rejected by the Council of Nice, A. D. 325. Cloth, 8vo., illustrated, $2.00.

THE PILGRIM'S PROGRESS, *as John Bunyan wrote it*. A fac-simile reproduction of the first edition, published in 1678. Antique cloth, 12mo., $1.25.

THE FAIREST OF THE FAIR, by Hildegarde Hawthorne. "The grand-daughter of Nathaniel Hawthorne possesses a full share of his wonderful genius." Cloth, 16mo., $1.25

A LOVER IN HOMESPUN, by F. Clifford Smith. Interesting tales of adventure and home life in Canada. Cloth. 12mo., 75 cents.

ANNIE BESANT: AN AUTOBIOGRAPHY. Cloth, 12mo., 368 pages, illustrated. $2.00

THE GRAMMAR OF PALMISTRY, by Katharine St. Hill Cloth, 12mo., illustrated, 75 cents.

AROUND THE WORLD IN EIGHTY MINUTES. Contains over 100 photographs of the most famous places and edifices with descriptive text. Cloth, 50 cents.

WHAT WOMEN SHOULD KNOW. A woman's book about women. By Mrs. E. B. Duffy. Cloth, 320 pages, 75 cents.

HENRY ALTEMUS' PUBLICATIONS.

THE CARE OF CHILDREN, by Elisabeth R. Scovil. "An excellent book of the most vital interest." Cloth, 12mo., $1.00.

PREPARATION FOR MOTHERHOOD, by Elisabeth R. Scovil. Cloth, 12mo., 320 pages, $1.00.

ALTEMUS' CONVERSATION DICTIONARIES. English-German, English-French. "Combined dictionaries and phrase books." Pocket size, each $1.00.

TAINE'S ENGLISH LITERATURE, translated from the French by Henry Van Laun, illustrated with 20 fine photogravure portraits. Best English library edition, four volumes, cloth, full gilt, octavo, per set, $10.00. Half calf, per set, $12.50. Cheaper edition, with frontispiece illustrations only, cloth, paper titles, per set $7.50.

SHAKESPEARE'S COMPLETE WORKS, with a biographical sketch by Mary Cowden Clark, embellished with 64 Boydell, and numerous other illustrations, four volumes, over 2000 pages. Half Morocco, 12mo., boxed, per set, $3.00.

DORE'S MASTERPIECES

THE DORE BIBLE GALLERY. A complete panorama of Bible History, containing 100 full-page engravings by Gustave Dore.

MILTON'S PARADISE LOST, with 50 full page engravings by Gustave Dore.

DANTE'S INFERNO, with 75 full page engravings by Gustave Dore.

DANTE'S PURGATORY AND PARADISE, with 60 full page engravings by Gustave Dore.

Cloth, ornamental, large quarto (9 x 12 inches), each $2.00.

TENNYSON'S IDYLLS OF THE KING, with 37 full page engravings by Gustave Dore. Cloth, full gilt, large imperial quarto (11 x 14½ inches), $4.50.

HENRY ALTEMUS' PUBLICATIONS.

THE RIME OF THE ANCIENT MARINER, by Samuel Taylor Coleridge, with 46 full page engravings by Gustave Dore. Cloth, full gilt, large imperial quarto (11 x 14½ inches), $3.00.

BUNYAN'S PILGRIM'S PROGRESS, with 100 engravings by Frederick Barnard and others. Cloth, small quarto (9 x 10 inches), $1.00.

DICKENS' CHILD'S HISTORY OF ENGLAND, with 75 fine engravings by famous artists. Cloth, small quarto, boxed (9 x 10 inches), $1.00.

BIBLE PICTURES AND STORIES, 100 full page engravings. Cloth, small quarto (7 x 9 inches), $1.00.

MY ODD LITTLE FOLK, some rhymes and verses about them, by Malcolm Douglass. Numerous original engravings. Cloth, small quarto (7 x 9), $1.00.

PAUL AND VIRGINIA, by Bernardin St. Pierre, with 125 engravings by Maurice Leloir. Cloth, small quarto (9 x 10), $1.00.

LIFE AND ADVENTURES OF ROBINSON CRUSOE, with 120 original engravings by Walter Paget. Cloth, octavo (7½ x 9¾), $1.50.

ALTEMUS' ILLUSTRATED LIBRARY OF STANDARD AUTHORS.

Cloth, Twelve Mo. Size, 5½ x 7¾ Inches. Each $1.00.

TALES FROM SHAKESPEARE, by Charles and Mary Lamb, with 155 illustrations by famous artists.

PAUL AND VIRGINIA, by Bernardin de St. Pierre, with 125 engravings by Maurice Leloir.

ALICE'S ADVENTURES IN WONDERLAND, AND THROUGH THE LOOKING-GLASS AND WHAT ALICE FOUND THERE, by Lewis Carroll. Complete in one volume with 92 engravings by John Tenniel.

LUCILE, by Owen Meredith, with numerous illustrations by George Du Maurier.

BLACK BEAUTY, by Anna Sewell, with nearly 50 original engravings.

SCARLET LETTER, by Nathaniel Hawthorne, with numerous original full-page and text illustrations.

THE HOUSE OF THE SEVEN GABLES, by Nathaniel Hawthorne, with numerous original full-page and text illustrations.

BATTLES OF THE WAR FOR INDEPENDENCE, by Prescott Holmes, with 70 illustrations.

BATTLES OF THE WAR FOR THE UNION, by Prescott Holmes, with 80 illustrations.

HENRY ALTEMUS' PUBLICATIONS.

ALTEMUS' YOUNG PEOPLES' LIBRARY

PRICE FIFTY CENTS EACH.

ROBINSON CRUSOE: (Chiefly in words of one syllable). His life and strange, surprising adventures, with 70 beautiful illustrations by Walter Paget.

ALICE'S ADVENTURES IN WONDERLAND, with 42 illustrations by John Tenniel. "The most delightful of children's stories. Elegant and delicious nonsense." —*Saturday Review.*

THROUGH THE LOOKING-GLASS AND WHAT ALICE FOUND THERE; a companion to "Alice in Wonderland," with 50 illustrations by John Tenniel.

BUNYAN'S PILGRIM'S PROGRESS, with 50 full page and text illustrations.

A CHILD'S STORY OF THE BIBLE, with 72 full page illustrations.

A CHILD'S LIFE OF CHRIST, with 49 illustrations. God has implanted in the infant heart a desire to hear of Jesus, and children are early attracted and sweetly riveted by the wonderful Story of the Master from the Manger to the Throne.

SWISS FAMILY ROBINSON, with 50 illustrations. The father of the family tells the tale of the vicissitudes through which he and his wife and children pass, the wonderful discoveries made and dangers encountered. The book is full of interest and instruction.

CHRISTOPHER COLUMBUS AND THE DISCOVERY OF AMERICA, with 70 illustrations. Every American boy and girl should be acquainted with the story of the life of the great discoverer, with its struggles, adventures, and trials.

THE STORY OF EXPLORATION AND DISCOVERY IN AFRICA, with 80 illustrations. Records the experiences of adventures and discoveries in developing the "Dark Continent," from the early days of Bruce and Mungo Park down to Livingstone and Stanley, and the heroes of our own times. No present can be more acceptable than such a volume as this, where courage, intrepidity, resource, and devotion are so admirably mingled.

HENRY ALTEMUS' PUBLICATIONS.

Altemus' Young Peoples' Library—continued.

THE FABLES OF ÆSOP. Compiled from the best accepted sources. With 62 illustrations. The fables of Æsop are among the very earliest compositions of this kind, and probably have never been surpassed for point and brevity.

GULLIVER'S TRAVELS. Adapted for young readers. With 50 illustrations.

MOTHER GOOSE'S RHYMES, JINGLES AND FAIRY TALES, with 234 illustrations.

LIVES OF THE PRESIDENTS OF THE UNITED STATES, by Prescott Holmes. With portraits of the Presidents and also of the unsuccessful candidates for the office; as well as the ablest of the Cabinet officers. It is just the book for intelligent boys, and it will help to make them intelligent and patriotic citizens.

THE STORY OF ADVENTURE IN THE FROZEN SEAS, with 70 illustrations. By Prescott Holmes. We have here brought together the records of the attempts to reach the North Pole. The book shows how much can be accomplished by steady perseverance and indomitable pluck.

ILLUSTRATED NATURAL HISTORY, by the Rev. J. G. Wood, with 8 illustrations. This author has done more to popularize the study of natural history than any other writer. The illustrations are striking and life-like.

A CHILD'S HISTORY OF ENGLAND, by Charles Dickens, with 50 illustrations. Tired of listening to his children memorize the twaddle of old fashioned English history the author covered the ground in his own peculiar and happy style for his own children's use. When the work was published its success was instantaneous.

BLACK BEAUTY, THE AUTOBIOGRAPHY OF A HORSE, by Anna Sewell, with 50 illustrations. A work sure to educate boys and girls to treat with kindness all members of the animal kingdom. Recognized as the greatest story of animal life extant.

THE ARABIAN NIGHTS ENTERTAINMENTS, with 130 illustrations. Contains the most favorably known of the stories.

HENRY ALTEMUS' PUBLICATIONS.

ALTEMUS' DEVOTIONAL SERIES.

Standard Religious Literature Appropriately Bound in Handy Volume Size. Each Volume contains Illuminated Title, Portrait of Author and Appropriate Illustrations.

WHITE VELLUM, SILVER AND MONOTINT, BOXED, EACH FIFTY CENTS.

1 KEPT FOR THE MASTER'S USE, by Frances Ridley Havergal. "Will perpetuate her name."

2 MY KING AND HIS SERVICE, OR DAILY THOUGHTS FOR THE KING'S CHILDREN, by Frances Ridley Havergal. "Simple, tender, gentle, and full of Christian love."

3 MY POINT OF VIEW. Selections from the works of Professor Henry Drummond.

4 OF THE IMITATION OF CHRIST, by Thomas A'Kempis. "With the exception of the Bible it is probably the book most read in Christian literature."

5 ADDRESSES, by Professor Henry Drummond. "Intelligent sympathy with the Christian's need."

6 NATURAL LAW IN THE SPIRITUAL WORLD, by Professor Henry Drummond. "A most notable book which has earned for the author a world-wide reputation."

7 ADDRESSES, by the Rev. Phillips Brooks. "Has exerted a marked influence over the rising generation."

8 ABIDE IN CHRIST. Thoughts on the Blessed Life of Fellowship with the Son of God. By the Rev. Andrew Murray. It cannot fail to stimulate and cheer.—*Spurgeon.*

9 LIKE CHRIST. Thoughts on the Blessed Life of Conformity to the Son of God. By the Rev. Andrew Murray. A sequel to "Abide in Christ." "May be read with comfort and edification by all."

10 WITH CHRIST IN THE SCHOOL OF PRAYER, by the Rev. Andrew Murray. "The best work on prayer in the language."

HENRY ALTEMUS' PUBLICATIONS.

11 **HOLY IN CHRIST.** Thoughts on the Calling of God's Children to be Holy as He is Holy. By the Rev. Andrew Murray. "This sacred theme is treated Scripturally and robustly without spurious sentimentalism."

12 **THE MANLINESS OF CHRIST**, by Thomas Hughes, author of "Tom Brown's School Days," etc. "Evidences of the sublimest courage and manliness in the boyhood, ministry, and in the last acts of Christ's life."

13 **ADDRESSES TO YOUNG MEN**, by the Rev. Henry Ward Beecher. Seven Addresses on common vices and their results.

14 **THE PATHWAY OF SAFETY**, by the Rt. Rev. Ashton Oxenden, D.D. Sound words of advice and encouragement on the text "What must I do to be saved?"

15 **THE CHRISTIAN LIFE**, by the Rt. Rev. Ashton Oxenden, D. D. A beautiful delineation of an ideal life from the conversion to the final reward.

16 **THE THRONE OF GRACE.** Before which the burdened soul may cast itself on the bosom of infinite love and enjoy in prayer "a peace which passeth all understanding."

17 **THE PATHWAY OF PROMISE**, by the author of "The Throne of Grace." Thoughts consolatory and encouraging to the Christian pilgrim as he journeys onward to his heavenly home.

18 **THE IMPREGNABLE ROCK OF HOLY SCRIPTURE**, by the Rt. Hon. William Ewart Gladstone, M P. The most masterly defence of the truths of the Bible extant. The author says: The Christian Faith and the Holy Scriptures arm us with the means of neutralizing and repelling the assaults of evil in and from ourselves.

19 **STEPS INTO THE BLESSED LIFE**, by the Rev. F. B. Meyer, B. A. A powerful help towards sanctification.

20 **THE MESSAGE OF PEACE**, by the Rev. Richard W. Church, D. D. Eight excellent sermons on the advent of the Babe of Bethlehem and his influence and effect on the world.

21 **JOHN PLOUGHMAN'S TALK**, by the Rev. Charles H. Spurgeon.

22 **JOHN PLOUGHMAN'S PICTURES**, by the Rev. Charles H. Spurgeon.

23 **THE CHANGED CROSS; AND OTHER RELIGIOUS POEMS.**

ALTEMUS' ETERNAL LIFE SERIES.

Selections from the writings of well-known religious authors, beautifully printed and daintily bound with original designs in silver and ink.

PRICE, 25 CENTS PER VOLUME.

1. ETERNAL LIFE, by Professor Henry Drummond.
2. LORD, TEACH US TO PRAY, by Rev. Andrew Murray.
3. GOD'S WORD AND GOD'S WORK, by Martin Luther.
4. FAITH, by Thomas Arnold.
5. THE CREATION STORY, by Honorable William E. Gladstone.
6. THE MESSAGE OF COMFORT, by Rt. Rev. Ashton Oxenden.
7. THE MESSAGE OF PEACE, by Rev R. W. Church.
8. THE LORD'S PRAYER AND THE TEN COMMANDMENTS, by Dean Stanley.
9. THE MEMOIRS OF JESUS, by Rev. Robert F. Horton.
10. HYMNS OF PRAISE AND GLADNESS, by Elisabeth R. Scovil
11. DIFFICULTIES, by Hannah Whitall Smith.
12. GAMBLERS AND GAMBLING, by Rev. Henry Ward Beecher.
13. HAVE FAITH IN GOD, by Rev. Andrew Murray.
14. TWELVE CAUSES OF DISHONESTY, by Rev. Henry Ward Beecher.
15. THE CHRIST IN WHOM CHRISTIANS BELIEVE, by Rt. Rev. Phillips Brooks.
16. IN MY NAME, by Rev. Andrew Murray.
17. SIX WARNINGS, by Rev. Henry Ward Beecher.
18. THE DUTY OF THE CHRISTIAN BUSINESSMAN, by Rt. Rev. Phillips Brooks.
19. POPULAR AMUSEMENTS, by Rev. Henry Ward Beecher.
20. TRUE LIBERTY, by Rt. Rev Phillips Brooks.
21. INDUSTRY AND IDLENESS, by Rev. Henry Ward Beecher
22. THE BEAUTY OF A LIFE OF SERVICE, by Rt. Rev. Phillips Brooks.
23. THE SECOND COMING OF OUR LORD, by Rev. A. T. Pierson, D D.
24. THOUGHT AND ACTION, by Rt. Rev. Phillips Brooks.
25. THE HEAVENLY VISION, by Rev. F. B. Meyer.
26. MORNING STRENGTH, by Elisabeth R. Scovil.
27. FOR THE QUIET HOUR, by Edith V. Bradt.
28. EVENING COMFORT, by Elisabeth R. Scovil
29. WORDS OF HELP FOR CHRISTIAN GIRLS, by Rev. F B. Meyer.
30. HOW TO STUDY THE BIBLE, by Rev. Dwight L. Moody.
31. EXPECTATION CORNER, by E. S. Elliot.
32. JESSICA'S FIRST PRAYER, by Hesba Stratton.

ALTEMUS' BELLES-LETTRES SERIES.

A collection of Essays and Addresses by eminent English and American Authors, beautifully printed and daintily bound, with original designs in silver.

PRICE, 25 CENTS PER VOLUME.

1. INDEPENDENCE DAY, by Rev. Edward E. Hale.
2. THE SCHOLAR IN POLITICS, by Hon. Richard Olney.
3. THE YOUNG MAN IN BUSINESS, by Edward W. Bok.
4. THE YOUNG MAN AND THE CHURCH, by Edward W. Bok.
5. THE SPOILS SYSTEM, by Hon. Carl Schurz.
6. CONVERSATION, by Thomas DeQuincey.
7. SWEETNESS AND LIGHT, by Matthew Arnold.
8. WORK, by John Ruskin.
9. NATURE AND ART, by Ralph Waldo Emerson.
10. THE USE AND MISUSE OF BOOKS, by Frederic Harrison.
11. THE MONROE DOCTRINE: ITS ORIGIN, MEANING AND APPLICATION, by Prof. John Bach McMaster (University of Pennsylvania).
12. THE DESTINY OF MAN, by Sir John Lubbock.
13. LOVE AND FRIENDSHIP, by Ralph Waldo Emerson.
14. RIP VAN WINKLE, by Washington Irving.
15. ART, POETRY AND MUSIC, by Sir John Lubbock.
16. THE CHOICE OF BOOKS, by Sir John Lubbock.
17. MANNERS, by Ralph Waldo Emerson.
18. CHARACTER, by Ralph Waldo Emerson.
19. THE LEGEND OF SLEEPY HOLLOW, by Washington Irving.
20. THE BEAUTIES OF NATURE, by Sir John Lubbock.
21. SELF RELIANCE, by Ralph Waldo Emerson.
22. THE DUTY OF HAPPINESS, by Sir John Lubbock.
23. SPIRITUAL LAWS, by Ralph Waldo Emerson.
24. OLD CHRISTMAS, by Washington Irving
25. HEALTH, WEALTH AND THE BLESSING OF FRIENDS, by Sir John Lubbock.
26. INTELLECT, by Ralph Waldo Emerson.
27. WHY AMERICANS DISLIKE ENGLAND, by Prof. Geo B Adams (Yale).
28. THE HIGHER EDUCATION AS A TRAINING FOR BUSINESS, by Prof. Harry Pratt Judson (University of Chicago).
29. MISS TOOSEY'S MISSION.
30. LADDIE.
31. J. COLE, by Emma Gellibrand.

HENRY ALTEMUS' PUBLICATIONS.

ALTEMUS' NEW ILLUSTRATED VADEMECUM SERIES.

Masterpieces of English and American Literature, Handy Volume Size, Large Type Editions. Each Volume Contains Illuminated Title Pages, and Portrait of Author and Numerous Engravings

Full Cloth, ivory finish, ornamental inlaid sides and back, boxed 40
Full White Vellum, full silver and monotint, boxed 50

1. CRANFORD, by Mrs. Gaskell.
2. A WINDOW IN THRUMS, by J. M. Barrie.
3. RAB AND HIS FRIENDS, MARJORIE FLEMING, ETC., by John Brown, M. D.
4. THE VICAR OF WAKEFIELD, by Oliver Goldsmith.
5. THE IDLE THOUGHTS OF AN IDLE FELLOW, by Jerome K. Jerome. " A book for an idle holiday."
6. TALES FROM SHAKSPEARE, by Charles and Mary Lamb, with an introduction by the Rev. Alfred Ainger, M. D.
7. SESAME AND LILIES, by John Ruskin. Three Lectures—I. Of the King's Treasures. II. Of Queen's Garden. III. Of the Mystery of Life.
8. THE ETHICS OF THE DUST, by John Ruskin. Ten lectures to little housewives on the elements of crystalization.
9. THE PLEASURES OF LIFE, by Sir John Lubbock. Complete in one volume.
10. THE SCARLET LETTER, by Nathaniel Hawthorne.
11. THE HOUSE OF THE SEVEN GABLES, by Nathaniel Hawthorne.
12. MOSSES FROM AN OLD MANSE, by Nathaniel Hawthorne.

HENRY ALTEMUS' PUBLICATIONS.

Altemus' New Illustrated Vademecum Series—continued.

13 TWICE TOLD TALES, by Nathaniel Hawthorne.

14 THE ESSAYS OF FRANCIS (LORD) BACON WITH MEMOIRS AND NOTES.

15 ESSAYS, First Series, by Ralph Waldo Emerson.

16 ESSAYS, Second Series, by Ralph Waldo Emerson.

17 REPRESENTATIVE MEN, by Ralph Waldo Emerson. Mental portraits each representing a class. 1. The Philosopher. 2. The Mystic. 3. The Skeptic. 4. The Poet. 5. The Man of the World. 6. The Writer.

18 THOUGHTS OF THE EMPEROR MARCUS AURELIUS ANTONINUS, translated by George Long.

19 THE DISCOURSES OF EPICTETUS WITH THE ENCHIRIDION, translated by George Long.

20 OF THE IMITATION OF CHRIST, by Thomas A'Kempis. Four books complete in one volume.

21 ADDRESSES, by Professor Henry Drummond. The Greatest Thing in the World; Pax Vobiscum; The Changed Life; How to Learn How; Dealing With Doubt; Preparation for Learning; What is a Christian; The Study of the Bible; A Talk on Books.

22 LETTERS, SENTENCES AND MAXIMS, by Lord Chesterfield. Masterpieces of good taste, good writing and good sense.

23 REVERIES OF A BACHELOR. A book of the heart. By Ik Marvel.

24 DREAM LIFE, by Ik Marvel. A companion to "Reveries of a Bachelor."

25 SARTOR RESARTUS, by Thomas Carlyle.

26 HEROES AND HERO WORSHIP, by Thomas Carlyle.

27 UNCLE TOM'S CABIN, by Harriet Beecher Stowe.

28 ESSAYS OF ELIA, by Charles Lamb.

HENRY ALTEMUS' PUBLICATIONS.

Altemus' New Illustrated Vademecum Series—continued.

29 MY POINT OF VIEW. Representative selections from the works of Professor Henry Drummond by William Shepard.

30 THE SKETCH BOOK, by Washington Irving. Complete.

31 KEPT FOR THE MASTER'S USE, by Frances Ridley Havergal.

32 LUCILE, by Owen Meredith.

33 LALLA ROOKH, by Thomas Moore.

34 THE LADY OF THE LAKE, by Sir Walter Scott.

35 MARMION, by Sir Walter Scott.

36 THE PRINCESS; AND MAUD, by Alfred (Lord) Tennyson.

37 CHILDE HAROLD'S PILGRIMAGE, by Lord Byron.

38 IDYLLS OF THE KING, by Alfred (Lord) Tennyson.

39 EVANGELINE, by Henry Wadsworth Longfellow.

40 VOICES OF THE NIGHT AND OTHER POEMS, by Henry Wadsworth Longfellow.

41 THE QUEEN OF THE AIR, by John Ruskin. A study of the Greek myths of cloud and storm.

42 THE BELFRY OF BRUGES AND OTHER POEMS, by Henry Wadsworth Longfellow.

43 POEMS, Volume I, by John Greenleaf Whittier.

44 POEMS, Volume II, by John Greenleaf Whittier.

HENRY ALTEMUS' PUBLICATIONS.

Altemus' New Illustrated Vademecum Series—continued.

45 THE RAVEN; AND OTHER POEMS, by Edgar Allan Poe.

46 THANATOPSIS; AND OTHER POEMS, by William Cullen Bryant.

47 THE LAST LEAF; AND OTHER POEMS, by Oliver Wendell Holmes.

48 THE HEROES OR GREEK FAIRY TALES, by Charles Kingsley.

49 A WONDER BOOK, by Nathaniel Hawthorne.

50 UNDINE, by de La Motte Fouque.

51 ADDRESSES, by the Rt. Rev. Phillips Brooks.

52 BALZAC'S SHORTER STORIES, by Honore de Balzac.

53 TWO YEARS BEFORE THE MAST, by Richard H. Dana, Jr.

54 BENJAMIN FRANKLIN. An Autobiography.

55 THE LAST ESSAYS OF ELIA, by Charles Lamb.

56 TOM BROWN'S SCHOOL DAYS, by Thomas Hughes.

57 WEIRD TALES, by Edgar Allan Poe.

58 THE CROWN OF WILD OLIVE, by John Ruskin. Three lectures on Work, Traffic and War.

59 NATURAL LAW IN THE SPIRITUAL WORLD, by Professor Henry Drummond.

60 ABBE CONSTANTIN, by Ludovic Halevy.

61 MANON LESCAUT, by Abbe Prevost.

HENRY ALTEMUS' PUBLICATIONS.

Altemus' New Illustrated Vademecum Series—continued.

62 THE ROMANCE OF A POOR YOUNG MAN, by Octave Feuillet.

63 BLACK BEAUTY, by Anna Sewell.

64 CAMILLE, by Alexander Dumas, Jr.

65 THE LIGHT OF ASIA, by Sir Edwin Arnold.

66 THE LAYS OF ANCIENT ROME, by Thomas Babington Macaulay.

67 THE CONFESSIONS OF AN ENGLISH OPIUM-EATER, by Thomas De Quincey.

68 TREASURE ISLAND, by Robert L. Stevenson.

69 CARMEN, by Prosper Merimee.

70 A SENTIMENTAL JOURNEY, by Laurence Sterne.

71 THE BLITHEDALE ROMANCE, by Nathaniel Hawthorne.

72 BAB BALLADS, AND SAVOY SONGS, by W. H. Gilbert.

73 FANCHON, THE CRICKET, by George Sand.

74 POEMS, by James Russell Lowell.

75 JOHN PLOUGHMAN'S TALK, by the Rev. Charles H. Spurgeon.

76 JOHN PLOUGHMAN'S PICTURES, by the Rev. Charles H. Spurgeon.

77 THE MANLINESS OF CHRIST, by Thomas Hughes.

78 ADDRESSES TO YOUNG MEN, by the Rev. Henry Ward Beecher.

79 THE AUTOCRAT OF THE BREAKFAST TABLE, by Oliver Wendell Holmes.

HENRY ALTEMUS' PUBLICATIONS.

Altemus' New Illustrated Vademecum Series—continued.

80 MULVANEY STORIES, by Rudyard Kipling.
81 BALLADS, by Rudyard Kipling.
82 MORNING THOUGHTS, by Frances Ridley Havergal.
83 TEN NIGHTS IN A BAR ROOM, by T. S. Arthur.
84 EVENING THOUGHTS, by Frances Ridley Havergal.
85 IN MEMORIAM, by Alfred (Lord) Tennyson.
86 COMING TO CHRIST, by Frances Ridley Havergal.
87 HOUSE OF THE WOLF, by Stanley Weyman.

AMERICAN POLITICS (non-Partisan), by Hon. Thomas V. Cooper. A history of all the Political Parties with their views and records on all important questions. All political platforms from the beginning to date. Great Speeches on Great Issues. Parliamentary Practice and tabulated history of chronological events. A library without this work is deficient. 8vo., 750 pages. Cloth, $3.00. Full Sheep Library style, $4.00.

NAMES FOR CHILDREN, by Elisabeth Robinson Scovil, author of "The Care of Children," "Preparation for Motherhood." In family life there is no question of greater weight or importance than naming the baby. The author gives much good advice and many suggestions on the subject. Cloth, 12mo., $ .40.

TRIF AND TRIXY, by John Habberton, author of "Helen's Babies." The story is replete with vivid and spirited scenes; and is incomparably the happiest and most delightful work Mr. Habberton has yet written. Cloth, 12mo., $ .50.

www.ingramcontent.com/pod-product-compliance
Lightning Source LLC
Chambersburg PA
CBHW032135230426
43672CB00011B/2342